Dislocations and Deformation Mechanisms in Thin Films and Small Structures

MATERIALS RESEARCH SOCIETY
SYMPOSIUM PROCEEDINGS VOLUME 673

Dislocations and Deformation Mechanisms in Thin Films and Small Structures

Symposium held April 17–19, 2001, San Francisco, California, U.S.A.

EDITORS:

Oliver Kraft

Max-Planck-Institut für Metallforschung
Stuttgart, Germany

Klaus W. Schwarz

IBM T.J. Watson Research Center
Yorktown Heights, New York, U.S.A.

Shefford P. Baker

Cornell University
Ithaca, New York, U.S.A.

L. Ben Freund

Brown University
Providence, Rhode Island, U.S.A.

Robert Hull

University of Virginia
Charlottesville, Virginia, U.S.A.

Materials Research Society
Warrendale, Pennsylvania

CAMBRIDGE
UNIVERSITY PRESS

University Printing House, Cambridge CB2 8BS, United Kingdom

One Liberty Plaza, 20th Floor, New York, NY 10006, USA

477 Williamstown Road, Port Melbourne, VIC 3207, Australia

314-321, 3rd Floor, Plot 3, Splendor Forum, Jasola District Centre, New Delhi - 110025, India

79 Anson Road, #06-04/06, Singapore 079906

Cambridge University Press is part of the University of Cambridge.

It furthers the University's mission by disseminating knowledge in the pursuit of education, learning and research at the highest international levels of excellence.

www.cambridge.org
Information on this title: www.cambridge.org/9781558996090

Materials Research Society
506 Keystone Drive, Warrendale, PA 15086
http://www.mrs.org

© Materials Research Society 2001

First published 2001
First paperback edition 2013

Single article reprints from this publication are available through University Microfilms Inc., 300 North Zeeb Road, Ann Arbor, MI 48106

CODEN: MRSPDH

A catalogue record for this publication is available from the British Library

ISBN 978-1-558-99609-0 Hardback
ISBN 978-1-107-41216-3 Paperback

CONTENTS

DISLOCATION AND DEFORMATION
MECHANISMS IN THIN METAL FILMS
AND MULTILAYERS I

DISCRETE DISLOCATIONS:
OBSERVATIONS AND SIMULATIONS

DISLOCATIONS AND DEFORMATION
MECHANISMS IN THIN FILMS AND
SMALL STRUCTURES

DISLOCATIONS IN
SMALL STRUCTURES

*Invited Paper

DISLOCATIONS AND DEFORMATION
IN EPITAXIAL LAYERS

DISLOCATION FUNDAMENTALS:
OBSERVATIONS, CALCULATIONS
AND SIMULATIONS

*Invited Paper

DISLOCATIONS AND DEFORMATION
MECHANISMS IN THIN METAL FILMS
AND MULTILAYERS II

PREFACE

It has been widely recognized that the mechanical properties of small volumes of materials such as thin films and patterned structures (lines, dots) can be very different from the mechanical properties of those same materials in bulk, and a number of MRS symposia have been dedicated to measurement and understanding of those properties. Many explanations of the mechanical behaviors of such small volumes have depended on simplified models of dislocation behavior. However, recent developments in dislocation modeling have made it possible to describe and understand dislocation behavior in much more detail than has previously been possible. Due to the increase in interest and capabilities in this area, this symposium, entitled "Dislocations and Deformation Mechanisms in Thin Films and Small Structures," held April 17–19 at the 2001 MRS Spring Meeting in San Francisco, California, was organized to focus on and discuss the special characteristics of dislocations in small volumes. The symposium was very well attended with 75 presentations from 15 countries divided into 6 oral sessions and 1 poster session. A wide range of topics was presented and discussed including mechanisms of plastic deformation in heteroepitaxial, multilayered, and polycrystalline thin films, as well as in 3D mesostructures such as epitaxial islands, semiconducting devices, and microcrystallites. It was our particular aim to stimulate an exchange between experimental work, theoretical modeling, and numerical simulations.

This volume contains a selection of the papers presented at this symposium. We hope that it will serve as a useful reference on deformation mechanisms and dislocations in thin films, and that it will stimulate new ideas and research in this exciting area of science and technology.

Oliver Kraft
Klaus W. Schwarz
Shefford P. Baker
L. Ben Freund
Robert Hull

July, 2001

ACKNOWLEDGMENTS

The success of this symposium is due to the efforts of many people. A note of thanks is extended to the speakers, poster presenters, and session chairs for their excellent presentations, discussions and critical reviews of the work included in this volume. We are also thankful to the MRS staff for their patience and capable assistance. We gratefully acknowledge the financial support of this symposium provided by:

Agere Systems
FEI Company
IBM T.J. Watson Research Center
JEOL USA, Inc.
MTS Systems Corp.
Nano Instruments Innovation Center

The efforts and generosity of these individuals has helped to enhance our state of knowledge in this scientifically interesting and technologically critical area.

MATERIALS RESEARCH SOCIETY SYMPOSIUM PROCEEDINGS

MATERIALS RESEARCH SOCIETY SYMPOSIUM PROCEEDINGS

Dislocation and Deformation Mechanisms in Thin Metal Films and Multilayers I

Mat. Res. Soc. Symp. Proc. Vol. 673 © 2001 Materials Research Society

Constrained Diffusional Creep in Thin Copper Films

D. Weiss, H. Gao, and E. Arzt
Max-Planck-Institut für Metallforschung and Institut für Metallkunde der Universität,
Seestr. 92, D-70174 Stuttgart, Germany

ABSTRACT

The mechanical properties of thin metal films have been investigated for many years. However, the underlying mechanisms are still not fully understood. In this paper we give an overview of our work on thermomechanical properties and microstructure evolution in pure Cu and dilute Cu-Al alloy films. Very clean films were produced by sputtering and annealing under ultra-high vacuum (UHV) conditions. We described stress-temperature curves of pure Cu films with a constrained diffusional creep model from the literature. In Cu-1at.%Al alloy films, Al surface segregation and oxidation led to a "self-passivating" effect. These films showed an increased high-temperature strength because of the suppression of constrained diffusional creep; however, under certain annealing conditions, these films deteriorated due to void growth at grain boundaries.

INTRODUCTION

For many years, materials research has focused on the understanding of deformation mechanisms in thin metal films on stiff substrates (see [1] and [2] for an overview). Copper and aluminum films on silicon substrates have been of special interest, as these materials are used today for microchip metallization. The large difference in thermal expansion coefficients of the film and substrate materials can lead to mechanical film stresses up to several 100 MPa, which are considered a serious reliability issue in microelectronic industry. Basic research has been interested in two specific phenomena: First, thin-film yield stresses at low temperatures are often proportional to the inverse of the film thickness. This effect has been explained in terms of a dimensional constraint on dislocation motion [1]. Second, the particular shape of a stress-temperature curve under thermal cycling depends not only on the film material, but can also change substantially if the film surface is protected by a thin protection layer (so called passivation) [3].

In this paper we report on the dramatic effect of vacuum conditions during film synthesis on the thermomechanical behavior and the microstructure of pure Cu films. Films produced under UHV conditions supported much higher stresses at high temperatures than conventional films produced under high-vacuum (HV) conditions. Stress-temperature curves of the UHV-produced Cu films could be well described with constrained diffusional creep. We furthermore report on alloying effects in Cu-1at.%Al films. Al surface segregation and oxidation led to a "self-passivating" effect. Stress-temperature curves were more "square" than the curves of pure Cu films and apparent signs of constrained diffusional creep were absent in these films. We also show that, under certain annealing conditions, large voids can grow at grain boundaries and grain-boundary triple junctions of the alloy films. A growth mechanism based on grain-boundary diffusional creep is discussed.

EXPERIMENT

Pure Cu films with thicknesses of 0.5 and 1 µm were produced by DC magnetron sputtering under UHV conditions (base pressure 1×10^{-8} Pa, 99.997% purity of Cu target). 2" Si wafers in (100) orientation, coated on both sides with 50 nm-thick amorphous SiO_x and SiN_x diffusion barriers, were used as substrates. Film synthesis, which is described in greater detail in [4], consisted of three steps: substrate cleaning by Ar ion bombardment, film deposition, and annealing (10 min at 600 °C). UHV conditions were maintained between all three steps. The sputtering gas (Ar) was further purified by a chemisorptive gettering system. Cu-1at.%Al films were produced in the same chamber by co-sputtering from a pure Al target (99.999% purity). After annealing, some films were oxidized at different temperatures and controlled oxygen pressures.

The microstructure of the films was analyzed with focused ion beam microscopy (FIB, FEI 200 workstation), transmission electron microscopy (TEM, JEOL 200CX), standard X-ray methods (Siemens D5000), Auger depth profiling, and X-ray photoelectron spectroscopy (XPS).

The mechanical film properties were measured during thermal cycling using the wafer curvature technique [5]. A wafer was placed on a tripod in a furnace continuously purged with nitrogen. Temperature was linearly increased from room temperature to either 500 or 600 °C at a rate of 6 K/min. Wafer curvature, which is proportional to film stress, was measured with a laser-scanning system.

RESULTS AND DISCUSSION

Mechanical behavior and microstructure of UHV-sputtered Cu and Cu-1at.%Al films

In Figure 1a, three stress-temperature curves of different 1 µm thick Cu films on Si substrates are compared. Both vacuum conditions during film synthesis and Al alloying have a large influence on the mechanical film properties. Thermal stress evolution in the pure Cu film, sputtered and annealed under UHV conditions, clearly differs from that in a Cu film fabricated at conventional HV conditions (Ref. [6]): the UHV Cu film supports significantly higher stresses than the HV film. Even higher stresses are found in the film alloyed with 1at.%Al. All these phenomena are due to different types of plastic deformation, which will be discussed below. Where thermal mismatch strain is accommodated mainly elastically, all curves are of similar shape. This occurs at the beginning of the heating cycle.

Vacuum conditions during film synthesis also had a large influence on the post-annealing microstructure. In contrast to HV Cu films known from the literature, which have a small grain size, mixed grain texture, and many twins [6], the UHV Cu films from this work had a larger grain size (median grain size 2.4 µm for a 1 µm thick film), very few twins (see Figure 1b), and a very strong and sharp {111} texture (see inset in Figure 1b).

The curve of the Cu-1at.%Al alloy film in Figure 1a looks similar to that of pure Al films known from the literature [7]. We found that surface segregation and oxidation of Al atoms during film oxidation led to a "self-passivating" effect, which is also described in the literature [8]. As we show in the following, the difference in high-temperature stresses between the Cu-Al alloy films and the pure Cu films can be quantitatively described by constrained diffusional creep, which is believed to be suppressed by the aluminum-oxide passivation in the alloy films.

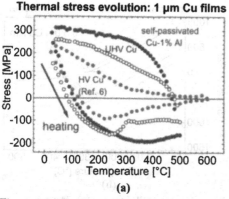

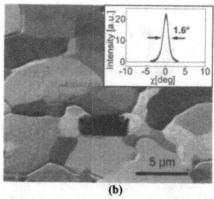

Thermal stress evolution: 1 μm Cu films

(a)

(b)

Figure 1 (a) Stress-temperature curves for 1 μm thick pure Cu and Cu-1at.%Al films (UHV, this work) and for a pure Cu film (HV, data taken from Keller et al., Ref. [6]). **(b)** FIB micrograph of 1 μm thick UHV-sputtered and annealed Cu film (with sputter-etched trench, sample tilt 45°). Inset: X-ray rocking curve of the {111} reflection of a similar film.

The constrained diffusional creep model

Grain-boundary diffusional creep in thin films has been the subject of many publications [9-12]. Recently, Gao and coworkers have presented a closed-form solution of constrained diffusional creep in films with columnar grains [13]. For the first time this problem has been solved for the constraint of perfect film adhesion, i.e. no sliding and no diffusion at the film/substrate interface was allowed. A schematic of this mechanism is shown in Figure 2a (top). Surface atoms, transported into the grain boundaries by directed diffusion under tensile stress, give rise to the formation of so called diffusion wedges at the grain boundaries: as a result of perfect film adhesion, the stress is relaxed more efficiently at the grain boundaries and at the film surface than in the grain interior and at the film/substrate interface.

Fast surface diffusion is required for constrained grain-boundary diffusional creep. The inhibition of surface diffusion would hence lead to an inhibition of this creep mechanism (Figure 2a, bottom). It is therefore plausible that constrained diffusional creep would be suppressed in the self-passivated Cu-Al alloy films. This also explains the similarity between the Cu-1at.%Al and Al films: Al is de facto passivated by the native oxide [3,14].

Figure 2b shows the theoretical stress evolution in a 0.5 μm thick Cu film, if constrained diffusional creep was the only plastic deformation mechanism. Similar results have been previously published by Dalbec et al. [15]. The average grain-boundary stress, shown as a dotted line, fully relaxes to zero at high temperatures. A similar curve would be obtained for the stress in a freestanding film, i.e. a film *without* substrate constraint. In contrast, the average stress in a film perfectly attached to the substrate (solid line in Figure 2b) does not relax to zero – even for fully relaxed grain boundaries.

A model for the stress evolution in Cu films should consider dislocation glide as well. We have recently suggested a modeling procedure schematically shown in Figure 3a [16]. Here, the reference stress σ_0, which is the stress in a film *without* diffusion wedges, is described by a

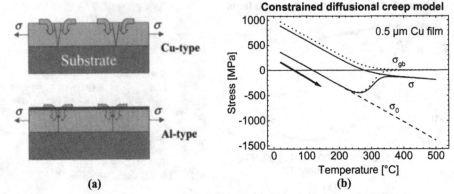

(a)   **(b)**

Figure 2 (a) Schematic for constrained diffusional creep under tensile stress (top). The grey arrows indicate atomic diffusion; the direction of diffusion would be reversed for compressive stress. A surface layer suppressing surface diffusion leads to the inhibition of constrained diffusional creep (bottom). **(b)** Modeled stress-temperature curve for a 0.5 μm thick Cu film with constrained diffusional creep as the only relaxation mechanism. σ_{gb} is the average grain boundary stress, σ_0 is a reference stress for a similar film without diffusion wedges, and σ is the average stress. The equations and model parameters are found in Ref. [16].

polynomial fit of the *experimental* stress σ_p of a self-passivated film of similar thickness. Grain-boundary stress and average stress are obtained numerically by solving the equations for constrained diffusional creep. In other words, we calculate the amount of stress which would be relaxed by constrained diffusional creep, if the passivation of the same film was removed. As a simplification, we neglected any influence of local stress relaxation by creep on dislocation glide. Figure 3b shows that the resulting average stress compares considerably well with the experimental data of an unpassivated film. As the only fitting parameter, grain size was chosen almost three times larger than the median grain size obtained by quantitative microstructure analysis. This larger effective grain size might be explained by unrelaxed low-energy boundaries.

Creep voiding in self-passivated Cu-1at.%Al films

In Cu-1at.%Al alloy films oxidized at or above 500 °C, large voids were found at many grain boundaries and grain-boundary triple junctions. Figure 4a shows a TEM image of a void at a triple junction in a Cu-1at.%Al film, which was annealed and subsequently oxidized at 600 °C in the sputtering chamber. A plausible growth mechanism is the diffusion of atoms from the surface of a void, nucleated at the intersection of a grain boundary with the surface oxide, into the grain boundary [17]. Driving force for void growth is the tensile film stress developing upon cooling. The inset in Figure 4a illustrates this mechanism, which is related to creep voiding in bulk alloys [18].

Figure 4b shows crack-like voids at grain boundaries and triple junctions in a similar film that was oxidized under less controlled conditions during thermal cycling to 600 °C in the wafer-curvature apparatus. Here, continuous oxidation of void surfaces during cooling caused the

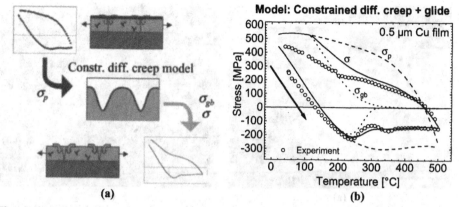

Figure 3 (a) Schematic for modeling constrained diffusional creep together with dislocation glide. σ_p is the experimental stress of a self-passivated film. (b) Model result in comparison with experimental data of a 0.5 µm thick unpassivated Cu film. A detailed description of the calculation is found in Ref. [16].

crack-like void shape. This particular void shape is also predicted for creep voids in bulk alloys, if diffusion on the void surface is rate limiting [18].

Grain-boundary voids were found neither in *unalloyed* Cu films, oxidized at 600 °C, nor in Cu-Al alloy films, annealed but *not oxidized*. Both systems exhibit high diffusivity at the film surface, which is why constrained diffusional creep, without the nucleation of voids, is energetically more favorable. Stress voiding can be a serious technological problem in thin conductor lines, where high tensile stresses may develop. Our results have unambiguously shown that this detrimental effect can be avoided in passivation-free films or by careful choice of the annealing parameters.

CONCLUSIONS

We have shown that synthesis of pure Cu films under very clean UHV conditions had large effects on both the thermomechanical behavior and the microstructure of these films. Stress-temperature curves displayed high residual stresses even at high temperatures. These curves were for the first time successfully described by a constrained diffusional creep model. Cu films alloyed with 1at.%Al supported even higher high-temperature stresses. Surface segregation and oxidation of Al atoms led to a self-passivating effect, which is believed to inhibit constrained diffusional creep. Cu-Al alloy films, annealed and oxidized at temperatures at or above 500 °C, were degraded due to voiding at grain boundaries and grain-boundary triple junctions. This detrimental effect can be avoided in passivation-free films or by careful choice of the annealing parameters.

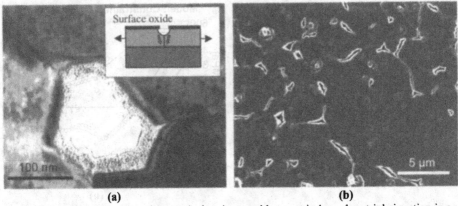

(a)	(b)

Figure 4 (a) TEM plan-view micrograph showing a void at a grain-boundary triple junction in a Cu-1at.%Al film, oxidized in the UHV sputtering chamber. Inset: Cross section of grain-boundary void (schematic). **(b)** FIB micrograph of crack-like voids in a 1 μm thick Cu-1at.%Al film, oxidized during thermal cycling to 600 °C in the furnace of the wafer curvature apparatus.

ACKNOWLEDGEMENTS

The authors acknowledge fruitful discussions with S. Baker, K.-N. Tu, O.S. Leung, W.D. Nix, M.J. Kobrinsky, and C.V. Thompson. This work was supported by the Deutsche Forschungsgemeinschaft under contract number AR 201/5.

REFERENCES

1. W. D. Nix, *Metall. Trans. A* **20A**, 2217-45 (1989).
2. E. Arzt, *Acta Mater.* **46**, 5611-26 (1998).
3. M. D. Thouless, K. P. Rodbell, and C. J. Cabral, *J. Vac. Sci. Tech. A* **14**, 2454-61 (1996).
4. D. Weiss, PhD Thesis, Universität Stuttgart (2000).
5. P. A. Flinn, D. S. Gardner, and W. D. Nix, *IEEE Trans. Electr. Dev.* **ED-34**, 689-99 (1987).
6. R.-M. Keller, S. P. Baker, and E. Arzt, *J. Mater. Res.* **13**, 1307-17 (1998).
7. Y.-C. Joo, P. Müllner, S. P. Baker, and E. Arzt, MRS Symp. Proc. Vol. 473, (Philadelphia, PA, 1997), p. 409-14.
8. J. Li, J. W. Mayer, and E. G. Colgan, *J. Appl. Phys.* **70**, 2820-7 (1991).
9. G. B. Gibbs, *Phil. Mag.* **13**, 589-93 (1966).
10. M. S. Jackson and C.-Y. Li, *Acta Met.* **30**, 1993-2000 (1982).
11. M. D. Thouless, *Acta Met.* **41**, 1057-64 (1993).
12. M. J. Kobrinsky and C. V. Thompson, *Appl. Phys. Lett.* **73**, 2429-31 (1998).
13. H. Gao, L. Zhang, W. D. Nix, C. V. Thompson, and E. Arzt, *Acta mater.* **47**, 2865-78 (1999).
14. R. P. Vinci, E. M. Zielinski, and J. C. Bravman, *Thin Solid Films* **262**, 142-53 (1995).
15. T. R. Dalbec, O. S. Leung, and W. D. Nix, in *Deformation, Processing, and Properties of Structural Materials*, edited by E. M. Taleff, C. K. Syn, and D. R. Lesuer (The Minerals, Metals & Materials Society, 2000), p. 95-108.
16. D. Weiss, H. Gao, and E. Arzt, *Acta mater., in press* (2001).
17. D. Weiss, O. Kraft, and E. Arzt, *Appl. Phys. Lett., submitted* (2001).
18. A. C. F. Cocks and M. F. Ashby, *Progr. Mat. Sci.* **27**, 189-244 (1982).

Mat. Res. Soc. Symp. Proc. Vol. 673 © 2001 Materials Research Society

An Experimental and Computational Study of the Elastic-Plastic Transition in Thin Films

Erica T. Lilleodden[1], Jonathan A. Zimmerman[2], Stephen M. Foiles[3] and William D. Nix[1]
[1]Department of Materials Science & Engineering, Stanford University, Stanford, CA 94305-2205, [2]Sandia National Laboratories, Livermore, CA 94551, [3]Sandia National Laboratories, Albuquerque, NM 87185

ABSTRACT

Nanoindentation studies of thin metal films have provided insight into the mechanisms of plasticity in small volumes, showing a strong dependence on the film thickness and grain size. It has been previously shown that an increased dislocation density can be manifested as an increase in the hardness or flow resistance of a material, as described by the Taylor relation [1]. However, when the indentation is confined to very small displacements, the observation can be quite the opposite; an elevated dislocation density can provide an easy mechanism for plasticity at relatively small loads, as contrasted with observations of near-theoretical shear stresses required to initiate dislocation activity in low-dislocation density materials. Experimental observations of the evolution of hardness with displacement show initially soft behavior in small-grained films and initially hard behavior in large-grained films. Furthermore, the small-grained films show immediate hardening, while the large grained films show the 'softening' indentation size effect (ISE) associated with strain gradient plasticity. Rationale for such behavior has been based on the availability of dislocation sources at the grain boundary for initiating plasticity. Embedded atom method (EAM) simulations of the initial stages of indentation substantiate this theory; the indentation response varies as expected when the proximity of the indenter to a Σ79 grain boundary is varied.

INTRODUCTION

Length-scale is fundamentally important to materials science, as evidenced by its relation to microstructure-property relations. For example, the flow stress of a material is known to scale with the square root of the dislocation density, inversely with the square root of the grain size, and, in the case of thin films, inversely with film thickness. However, a single-valued flow stress, considered to be a material property, is typically used in continuum mechanics. This removes any explicit length-scale dependence from the constitutive relations, taking into account only the average microstructure of the material. This is a reasonable approach in cases where the scale of deformation is large relative to the scale of microstructural inhomogeneities. However, as the characteristic length-scale of the deformation field tends toward the characteristic material length-scale, the governing relations between stress and strain may deviate from classical laws. Indeed, anomalous yielding behavior has been commonly observed in nanoindentation studies. Discrete load-displacement response in the early stages of indentation (e.g. several nanometers to tens of nanometers) has been widely observed [e.g. 2, 3, 4], and can be associated with the nucleation of dislocations. At a larger length-scale (>100nm) the observation of decreasing hardness with increasing displacement can be explained by strain gradient plasticity [e.g. 1]. However, both of these anomalies are most easily discussed without explicitly considering the microstructure of the indented materials. It is thus the objective of this work to explore the effect of microstructure on the initiation and evolution of plasticity.

EXPERIMENTAL

The experimental work was accomplished with a 1 micron gold film evaporated onto a Si (001) substrate, and a 0.16 micron thick gold film sputtered onto an amorphous silica substrate. Both films showed strong (111) texture. The silica substrate was used for the thinner film in order to minimize substrate effects, since silica and gold have similar elastic properties. The absence of a native oxide layer makes gold an ideal system for studying the onset of plasticity, since the absence of an oxide removes the possibility of oxide fracture or other oxide/metal interface defect effects. The films were annealed in order to grow the grains and to remove any pre-existing dislocations from the grain interiors. A 3 to 4nm W layer was deposited on the bare substrates, to promote adhesion of the films. Atomic Force Microscopy (AFM) was used to characterize the grain size and surface roughness of the samples. The 1 micron thick film had a grain size roughly 2 to 3 times the film thickness, while the 0.16 micron thick film had a grain size on the order of the film thickness. Nanoindentation experiments were conducted with a Nanoindenter XP with a DCM head (MTS, Minneapolis, MN). This instrument provides a continuous measurement of the contact stiffness via a superimposed oscillation of the load during loading. The continuous stiffness measurement (CSM) technique [5] greatly improves the function of nanoindentation experiments, by offering continuous measurement of the elastic and plastic response of the material during loading, rather than relying a set of partial unloading curves and/or multiple indentations for valid characterization. This technique is critical to understanding the evolution of plasticity during a single indentation, which is important in understanding the dependence of plastic deformation on local microstructural environment.

RESULTS AND DISCUSSION

Figure 1 compares the indentation behavior of the small grained 0.16 micron thick Au/silica sample and the large grained 1 micron Au/Si sample. Several distinct differences are clearly observed. The 1 micron thick film shows elastic discrete load-displacement behavior and larger sustained loads in the initial stages of deformation. In contrast, the small grained film shows continuous loading, with no discrete transition between elastic and plastic deformation. The hardness was measured in the conventional way, using a calibrated indenter-tip area function for the calculation of the contact area. The difference in hardness-displacement behavior between the two samples is particularly striking. The development of the mean pressure (defined as the indentation hardness, but not necessarily associated with plastic deformation) in the 1 micron thick film during elastic loading reaches a value of 2.2 GPa, whereas the polycrystalline film shows initially soft behavior and a hardness of 0.5 GPa. Following the first pop-in during indentation into the 1 micron thick film, the hardness is observed to decrease, as may be ascribed to continued dislocation nucleation or strain gradient plasticity effects. In contrast, the small grained film shows increased hardness with increasing displacement. The increased hardness can be attributed to conventional Hall-Petch behavior upon inspection of the number of grains contacted during loading, as shown in Figure 1(d). Dividing the contact area by the average grain size reveals multigrain contact in the case of the 0.16 micron thick film, while the 1 micron film is well described by single crystal behavior, within the first 100nm of displacement.

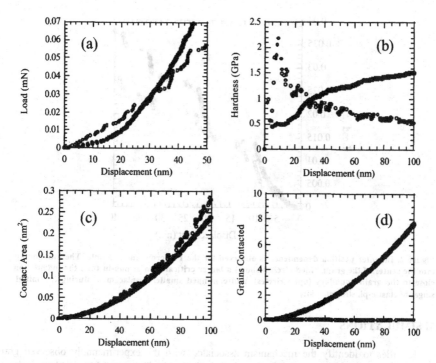

Figure 1. (a) Load and (b) hardness behavior of the 1 micron thick Au film on Si (open circles), and the 160nm small grained Au film on silica (filled circles), show significant differences. Assessment of the (c) contact area and subsequently (d) the number of grains contacted during indentation reveal a Hall-Petch type hardening mechanism in the small grained sample, while the large grained film shows the nucleation and strain gradient effects associated with single crystal behavior.

The differing behavior of the two films can be rationalized in terms of the availability of dislocation sources. Grain boundaries may act as effective dislocation sources, rather than requiring high energy-cost nucleation events to initiate plasticity. Thus, indentations into volumes that include or are near a grain boundary may allow easy initiation of plasticity. Consistent with this theory, the small grained film shows continuum elastoplastic behavior immediately, in contrast to the discrete elastic-plastic transitions observed in the large grained film. Additionally, the small grained film showed reproducible load-displacement behavior for random indentation positions across the sample. This is in sharp contrast to the load-displacement sensitivity to grain boundary proximity for the larger grained 1 micron thick film. Figure 2 shows the load-displacement behavior of two indentations: one indent positioned in the center of a grain, and another indent positioned close to the grain boundary. It is revealed that the indentation into the center of the grain leads to a higher critical load for pop-in and smaller overall displacements during the initial stages of plastic deformation, as compared with the indentation near the grain boundary.

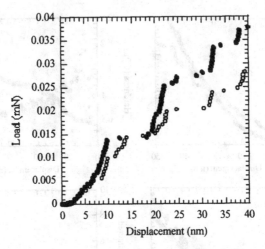

Figure 2. Indenter position dependence is observed for the 1 micron Au/Si sample. The indentation into the center of the grain (filled circles) shows a larger critical load at pop-in than the indentation close to the grain boundary (open circles), and continued smaller displacement during the initial stages of elastic-plastic behavior.

EAM SIMULATIONS

In order to identify the mechanism associated with the experimentally observed grain boundary proximity effect the embedded atom method (EAM) [7] was used to simulate the load-displacement response near and away from a $\Sigma 79$ grain boundary in gold. A 237,747 atom system was considered, with fixed boundaries along x = +/- 150 Å and z = -100 Å (simulating an infinitely stiff substrate), a free surface (to be indented) at z = 0, and a periodic boundary at y = +/- 67 Å. The normal to the grain boundary plane is in the x direction. A repulsive potential spherical indenter was used with a 40 Å radius of curvature. Indentation was simulated by increasing the displacement of the indenter in 0.1 Å steps, allowing the system to minimize its energy, and then evaluating the applied load. Three initial positions of the indenter relative to the grain boundary were chosen: 50 Å from the boundary, 25 Å from the boundary, and 15 Å from the boundary. The load-displacement response associated with these three indenter positions is shown in Figure 3a. At a critical load dislocation nucleation occurs, as signified by a discrete drop in the load. This is analogous to the displacement bursts during load-controlled experiments on the 1 micron thick film. The indentation 50 Å from the grain boundary resulted in the nucleation of a dislocation dissociated into 2 sets of partial dislocations on two adjacent {111} planes, as has been observed in simulations of indentation into perfect crystals [8]. The indentation 15 Å from the grain boundary results in the emission a perfect dislocation from the grain boundary. The dislocation slipped across a $(11\overline{1})$ plane, and cross-slipped onto the adjoined $(\overline{1}11)$ plane. This is shown in Figure 3b, a 3-D perspective of the post load-drop configuration underneath the contact. A slip vector analysis [9] was used to determine the Burgers vector associated with the deformation, and was found to be $\frac{1}{2}\langle\overline{1}0\overline{1}\rangle$. In order to identify the mechanism of grain boundary emission, a dynamic simulation was accomplished during the load drop. Figure 4 shows four time steps during this dynamic simulation, identifying high-energy atoms with a darker color.

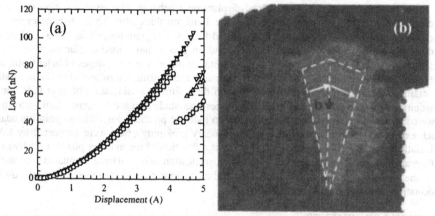

Figure 3. (a) EAM load-displacement response for 3 indenter positions: 50 Å away from the grain boundary (inverted triangles), 25 Å from the grain boundary (upright triangles), and 15 Å from the grain boundary (circles). The critical load is reduced as the indenter position approaches the grain boundary. (b) 3-D image underneath the contact shows the slip planes (designated by dashed lines) and the Burgers vector. Long arrows show the direction of slip.

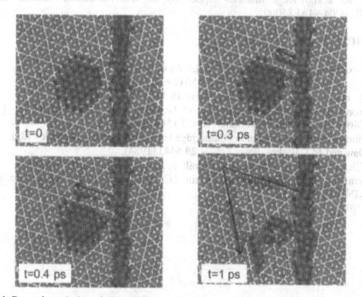

Figure 4. Dynamic evolution of dislocation emission from the grain boundary is shown. Dark gray atoms indicate the strained contact region, the grain boundary, and the dislocation line. The black arrows show the direction and extent of slip.

SUMMARY

It has been shown that indentation size effects are observed and that they are dependent on the local microstructure. Discrete load-displacement behavior implies dislocation nucleation and/or emission events; this was observed in the 1 micron thick film. These discrete events are not experimentally observed in the small grained film, where grain boundaries act as dislocation sources for easy initiation of plasticity and result in a continuous load-displacement response. Additionally, multiple-grain contact during indentation during the early stages of indentation into the small grained film lead to a Hall-Petch hardening mechanism that overwhelms the nucleation or strain gradient ISE observed in the 1 micron thick film. Consideration of easy initiation of plasticity due to the presence of grain boundaries was studied in the 1 micron film, showing a lowered critical load for discrete response for an indenter position closer to the grain boundary. Such experimental observations of a grain boundary proximity effect were supported by EAM calculations of the load-displacement response as a function of the indenter position relative to a Σ79 grain boundary. Homogeneous dislocation nucleation occurs during simulated indentation into the perfect crystal, while dislocation emission from the grain boundary occurs during indentation close to the grain boundary.

ACKNOWLEDGEMENTS

ETL and WDN acknowledge financial support from the Division of Materials Sciences of the Office of Basic Energy Sciences of the DOE under Grant No. DE-FG03-89ER45387-A008. JAZ and SMF acknowledge financial support by the DOE at Sandia National Labs under contract DE-AC04-94AL85000.

REFERENCES

1. W.D. Nix and H. Gao, J. Mech. Phys. Solids **46** 411 (1998).
2. T.F. Page, W.C. Oliver, J. Mater. Res. **7** 450 (1992).
3. T.A. Michalske and J.E. Houston, Acta Mat. **46** 391 (1998).
4. S. Corcoran, R. Colton, E. Lilleodden and W. Gerberich, Phys. Rev. B **55** R16057 (1997).
5. W.C. Oliver and G.M. Pharr, J. Mater. Res. **7** 1564 (1992).
6. K.L. Johnson, *Contact Mechanics* (Cambridge University press, New York 1985).
7. M.S. Daw and M.I. Baskes, Phys. Rev. B **29** 6443 (1984).
8. C.L. Kelchner, S.J. Plimpton and J.C. Hamilton, Phys. Rev. B **58** 11085 (1998).
9. J.A. Zimmerman, C.L. Kelchner, P.A. Klein, J.C. Hamilton and S.M. Foiles, Phys. Rev. Let., (2001) IN REVIEW.

Mat. Res. Soc. Symp. Proc. Vol. 673 © 2001 Materials Research Society

"REVERSE" STRESS RELAXATION IN CU THIN FILMS

R. Spolenak[1], C. A. Volkert[2], S. Ziegler[1,3], C. Panofen[1,3], W.L. Brown[1]

[1] Agere Systems, formerly of Bell Laboratories, Lucent Technologies, Murray Hill, NJ 07974 USA
[2] Max-Planck-Institut für Metallforschung, Stuttgart, Germany
[3] I. Physik. Inst. A., Lehrstuhl für Physik neuer Materialien, Aachen, Germany

ABSTRACT

In this study we present investigation on the anelastic behavior of sputtered 1 μm thin Cu films. Most of the literature that reports on the mechanical properties of thin metallic films is based on substrate curvature measurements. We have developed a new version of a bulge tester that combines the capacitive measurement of the bulge deflection of a membrane with a resonance frequency measurement of the residual stress in the membrane. A Cu membrane is plastically deformed to a pre-determined strain by controlled gas-pressure bulging of the membrane. After the bulging stress is removed, the residual tensile stress, which has been decreased by the plastic deformation, is then determined by measuring the resonant frequency as a function of time. Immediately after plastic straining, the residual (tensile) stress of membranes was observed to **increase**. At room temperature a maximum stress was typically reached in the order of an hour. At still longer times the stress decreased again as a result of creep. The transient increase in stress following plastic straining grew larger as the amount of plastic strain produced by bulging was increased. With higher temperatures the transient became both faster and larger. A model is presented that based on the mechanism of thermally activated glide separates the microstructure in a class of "soft" and "hard" grains solving the issue of an "apparent" increase in strain energy as a function of time after deformation.

INTRODUCTION

The understanding of mechanical stress in thin metallic films is of essential importance for the reliability of microdevices. The applications for these devices range from metallizations for contacts to reflective coatings in MEMS devices. In some cases metals are even used for structural components. In general two sources of stress are present in these films. One is the stress caused by the method of deposition and the second is stress induced by the mismatch in thermal expansion coefficients between the metal and the (usually) semiconducting substrate.

For reliability purpose, it is essential that these stresses do not change over time and do not lead to local deformation as, for instance, stress voiding. In general it is important to understand the interaction between stresses on a local level and the macroscopic changes that they cause.

In this study, we present an effect that macroscopically can be interpreted as an anelastic (a time dependent, but reversible) effect. We find that thin Cu films show an increase in tensile stress over time after having been plastically deformed. Similar observations have been made by Vinci et al. [1] on free standing Al thin films in a microtensile apparatus. In the present case the method of the deformation was a bulge test [2-4] and subsequent measurements of changes in stress were made by monitoring the resonant frequency of a membrane [5,6]. The unique combination of the two techniques in one tool [4] made this experiment possible.

The time behavior of deformation was found to be logarithmic, a phenomenon that has been described as exhaustion creep [7]. In bulk material the structural behavior is known as thermally activated glide [8] and has recently been applied to thin films [9-12].

In our case, however, we observe an "inverse" stress relaxation, which has the opposite sign to the deformation mechanisms described above. The strain energy of the system appears to be increasing over time following removal of a deforming stress. Such a physically provocative conclusion has led us to an explanation that depends upon microscopic variations in the deformation behavior of different grains in the film as described below.

EXPERIMENTAL

The samples investigated in this work were 1 micron thick sputtered films on Si rich SiN_x (200 nm). The SiN_x (deposited on both sides of the Si wafers) was designed to have a residual stress of about 100 MPa in tension. After metal deposition the samples were annealed for 30 min at 400 °C in forming gas. Subsequently, rectangular membranes (3x12 mm^2) were etched from the backside of the Si wafer using hot KOH. The SiN_x on the back of the wafer was patterned with RIE and served as the mask for defining the etch window. The SiN_x on the front side of the wafer (under the Cu) served as an etch stop when the KOH had removed all of the Si in the window. Details of the process can be found elsewhere [4].

A 20x20 mm^2 chip in which the membrane was centered was O-ring sealed over a hole in a small pressure chamber, which in turn was immersed in a vacuum chamber. The gas (Ar) pressure in the inner chamber was increased and the membrane bulged. The deflection of the bulge was determined capacitively. From the Ar pressure and the bulge height, stress-strain curves could be computed. Due to the rectangular shape of the membranes a uniaxial stress state was superimposed on the initial biaxial residual stress in the membrane.

After bulging, the inner chamber was evacuated and the membrane was resonantly excited electrostatically. Its resonant frequency allowed determination of residual stress with high sensitivity (see results). The residual stress was monitored as a function of time. The apparatus was temperature stabilized to < 0.5K to remove the effect of temperature fluctuations on the membrane stress. Further details of the tool can be found in Spolenak et al. [4].

The measurements described above were carried out following different total strains ranging from 0.8 to 2 x10^{-3} and at temperatures ranging from 30 °C to 125 °C.

RESULTS

Fig. 1 shows two bulge tests. After the first test, which shows a significant amount of plastic deformation, both chambers described above were evacuated and the resonant frequency of the membrane was monitored as a function of time. Subsequently, it was converted into stress using the membrane dimensions, the mode number of the resonance and the average density of the membrane as input data [5]. The difference in residual stress between the end of the first bulge test and the beginning of the second can be attributed to "reverse" stress relaxation.

The time behavior of this relaxation can be seen in Fig. 2. The curve can be fitted by a logarithmic function whose origin will be discussed later. As one can see, the function fits the data nicely leading to three fit parameters.

The results for several experiments on a Cu only membrane and a Cu film on SiN_x can be found in Table 1. With an increasing amount of plastic deformation (higher pressures) the residual stress is decreased as described by parameter A. Parameter C is a time offset as the "actual" start of the experiment cannot be accurately determined.

Parameter B describes the time dependence of the phenomena. Fig. 3 shows the dependence of parameter B on the total strain that the sample had seen in the prior bulge experiment. Parameter B increases linearly with the total strain. A minimum strain of about 0.9x10^{-3} is needed to activate the process. This is true for both the Cu only film as well as the Cu film on SiN_x. The major difference between the two is the stronger apparent decrease in residual stress for the Cu only film, which is masked by the purely elastic behavior of SiN_x for Cu on SiN_x. The presence of SiN_x does not change the time dependence of the phenomena (parameter B) as a function of the total strain. However, a higher pressure is needed during bulging to achieve the same degree of deformation.

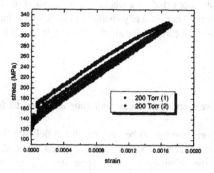

Figure 1: two bulge tests for Cu on SiN_x up to 200 Torr

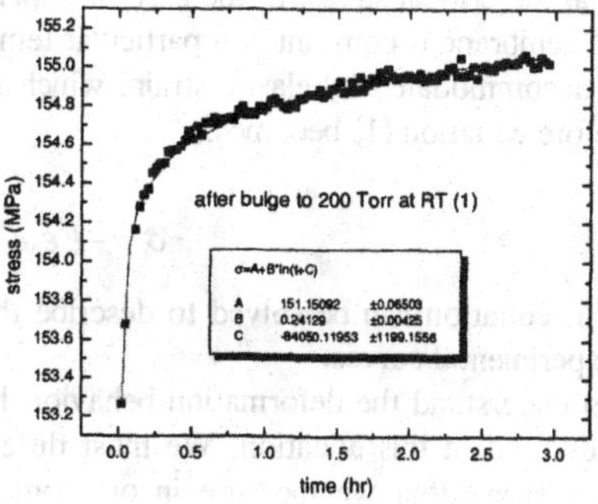

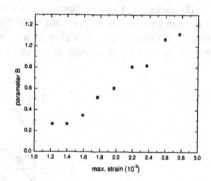

Figure 2: A "reverse" stress relaxation curve with logarithmic fit.

	A (MPa)	B	C (ms)	ε_{max} (10^{-3})
160 Torr	173.7	0.265	28000	1.22
180 Torr	163.3	0.265	-334000	1.40
200 Torr	149.5	0.347	-314000	1.59
220 Torr	133.9	0.518	-230000	1.77
240 Torr	118.2	0.605	-286000	1.97
260 Torr	97.8	0.807	-12000	2.19
280 Torr	78.4	0.818	-346000	2.37
300 Torr	61.0	1.064	-180000	2.59
320 Torr	43.7	1.115	-120000	2.77

Table 1: Fitting parameters for eq. 6 (Cu at RT)

Figure 3: The dependence of parameter B in the fit to the maximum

DISCUSSION AND CONCLUSIONS

These surprising observations, in which the energy of the macroscopic system seems to increase with time, can be explained by considering microplasticity on a microscopic level. As one has seen in the results section, the measurements can be very well fitted by a model of thermally activated dislocation glide [8]:

$$\dot{\varepsilon}_{pl} = \dot{\varepsilon}_0 e^{\frac{-\Delta F}{k_b T}\left(1 - \frac{\sigma}{\sigma_{crit}}\right)} \tag{1}$$

where σ_{crit} is the critical stress at which dislocations are able to move past an obstacle without temperature assistance, ΔF is the activation energy of the process when the stress is less than the critical stress and temperature assistance enhances the motion and $(d\varepsilon/dt)_0$ is the rate of change of strain at the critical stress. In the case of a membrane supported by a silicon frame the total strain in the membrane is constant at a particular temperature. Therefore any change in plastic strain has to be accommodated by elastic strain, which in our case can be measured as a change in stress. Therefore equation (1) becomes:

$$\dot{\sigma} = -E\dot{\varepsilon}_0 e^{\frac{-\Delta F}{k_b T}\left(1 - \frac{\sigma}{\sigma_{crit}}\right)} \tag{2}$$

This equation can be solved to describe the development of stress over time and used to fit our experimental curves.

To understand the deformation behavior, however, one has to be very careful in defining the stresses used in this equation. We must differentiate between the macroscopic stress level, the average stress that we measure in our tool, and the microscopic local stress that causes the movement of dislocations and thus leads to a macroscopically visible phenomenon.

Since on a macroscopic level the strain energy of the system is increasing, one could assume that this increase in energy could be compensated by reduction of energy by reducing the number of defects in the system. On an atomistic level, however, it is hard to see why a dislocation should move in the opposite direction to the force that acts on it for several Burgers vectors to annihilate at an interface.

The explanation that we have found for our observations is based on differentiating the grains in the film into two classes; (a) hard grains that do not show any plastic deformation, which is the majority component and (b) a class of soft grains that deform plastically even if the whole of the thin film seems to be deforming almost elastically (see Fig. 4).

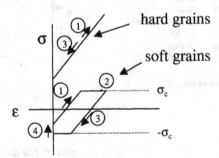

Figure 4: schematic of deformation of thin Cu membrane: 1: bulging, 2: constant stress test, 3: unloading, 4: "reverse" stress relaxation

The soft grains will see a significant degree of plastic deformation during loading and will end up in a compressive stress state upon unloading. The idea of separating the grains into two classes is based on the experimental evidence that even in an on-average tensile film compressive grains can be found in sputtered thin film Cu [13]. The total stress σ_m in the film is then given by:

$$\sigma_m = f_t\sigma_t + f_c\sigma_c \tag{3}$$

where t and c denominate the tensile stress of the hard grains and the compressive stress of the soft grains after the loading cycle shown above. The f's are their corresponding volume fractions.

Assuming that the class of hard, tensile grains do not deform over the time scale investigated, the local stress leading to macroscopic deformation is determined by changes in the soft, compressive grain component. This solves two problems:

- The force on dislocations in the soft grains leads to a reduction in their compressive stress. Dislocations move in the direction given by the force acting on them. The reduction of compressive stress of the soft grains increases the median stress of the film and makes it more tensile on a macroscopic scale.

- The apparent increase in the strain energy of the macroscopic system does not take account of the microscopic details. The total strain energy of the system is given by the sum of the strain energies of every single grain. In this summation only the absolute values of stress are relevant. Therefore, by reducing the compressive stress in the class of soft grains the total strain energy actually is reduced, even though this leads to an increase in macroscopic tensile strain.

In this way the change in macroscopic stress σ_m is given by the change of stress in the class of compressive grains. This straight-forward assumption is based on a parallel arrangement of hard and soft grains. In real polycrystal the configuration is more complicated, but for a basic understanding this simple picture suffices. Then equation 2 becomes:

$$\dot{\sigma}_m = f_c \dot{\sigma}_c = -f_c E \dot{\varepsilon}_0 e^{\frac{-\Delta F}{k_b T}\left(1 - \frac{\sigma_c}{\sigma_{crit}}\right)} \tag{4}$$

As we need to fit it to our macroscopic data we shall write equation 4 in terms of σ_m:

$$\dot{\sigma}_m = f_c \dot{\sigma}_c = -f_c E \dot{\varepsilon}_0 e^{\frac{-\Delta F}{k_b T}\left(\left(1 + \frac{f_t}{f_c}\frac{\sigma_t}{\sigma_{crit}}\right) - \frac{\sigma_m}{f_c \sigma_{crit}}\right)} \tag{5}$$

The curves in the results section have been fitted to:

$$\sigma = A + B \ln(t + C) \tag{6}$$

where the parameters are given by:

$$A = f_t \sigma_t + f_c \sigma_c - \frac{k_b T \sigma_{crit} f_c}{\Delta F} \ln(-\frac{\Delta F}{k_b T}\frac{E \dot{\varepsilon}_0}{\sigma_{crit}})$$

$$B = -\frac{k_b T \sigma_{crit} f_c}{\Delta F} \tag{7}$$

$C \ldots$ time constant

In parameter A the dominant term is the first (all the others are much smaller) reflecting the macroscopic residual stress. The parameter B reflects the amount of soft grains. As one can see in the results section, B increases with absolute strain as well as with temperature. The other parameters should be constant assuming that the microstructure does not change during deformation. This has been verified by bulging to the same pressure several times and always finding the same results. If one assumes a distribution of compressive yield stresses σ_{crit} rather than a single value, it becomes obvious that the volume fraction of compressive grains f_c has to increase as a function of total strain. In addition one observes a threshold for parameter B which is indicative of a region of total strain in which the entire film behaves plastically.

As σ_{crit} is compressive, it is negative making all the fitting parameters positive. One can see the difficulty in determining the activation energy for the process through this kind of fit as no parameter is strongly dependent on it. However, its order of magnitude can be evaluated from the absolute value of parameter B and is reasonable.

Parameter C describes the time offset of the process. The beginning of the process cannot be accurately determined. However, the time offset should increase with the speed of the process and in fact this is observed when one increases the temperature.

In general the next question to consider is why two classes of grains exist. The yield stress of a single grain depends on a variety of factors. The most prominent ones are grain size, defect density and crystal orientation with respect to the applied stress state. The latter one is not only dependent on a particular grain, but also on its next neighbors. The influence of a neighboring grain extends as far as the thickness of the film. Since in a stabilized metallic, polycrystalline thin film the grain size is of the order of the film thickness, the influence of interaction of stresses between the next neighbors can be significant. As the sputtered Cu films contain a fraction of (100) textured grains one could envision that those could form the fraction of "soft" grains. However, microdiffraction experiments [13] show that the compressive stresses are found in (111) textured grains. This further substantiates the importance of grain to grain interaction.

It will be the focus of ongoing research to investigate, what kind of interactions between next neighbors and what crystallographic configuration is necessary to make a grain "soft" or "hard".

ACKNOWLEDGEMENTS

The authors would like to thank the members of the SFRL (Silicon Fabrication Research Laboratory) namely, M. Buonanno, M. Hoover, S. Fiorillo, M. D. Morris, T. Craddock, W. Mansfield and furthermore M. Peabody, J. Li, C. Caminos and G. Bogart for their contributions in preparation of the samples.

REFERENCES

1. R. P. Vinci, G. Cornella, and J. C. Bravman, AIP Conf. Proc. **491**, 240-248 (1999).
2. A. J. Kalkman, A. H. Verbruggen, G. C. A. M. Janssen, and F. H. Groen, Rev. of. Sci. Inst. **70**(10), 4026-4031, (1999).
3. J. Vlassak, and W. D. Nix, J. Mater. Res. **7**, 1553 (1992).
4. R. Spolenak, S. Ziegler, C. A. Volkert, B. Boie, J. Kraus, D. Kossives, and W. L. Brown, in preparation (2001).
5. M. P. Schlax, R. L. Engelstad, E. G. Lovell, J. A. Liddle; and A. E. Novembre, Proc. of the SPIE, **3676**, 152-161 (1999).
6. M. P. Schlax, R. L. Engelstad, E. G. Lovell, J. A. Liddle; and A. E. Novembre, J. of Vac. Sci. & Tech. B, **17**(6), 2714-18 (1999).
7. A.H. Cottrell, "Dislocations and Plastic Flow in Crystals", Oxford University Press, London (1953)
8. H. J. Frost and M. F. Ashby, "Deformation mechanism maps", Pergamon Press, Oxford (1982).
9. P. A. Flinn, D. S. Gardner, and W. D. Nix, IEEE Trans. Electr. Dev., ED-34(3), 689-99
10. M. D. Thouless, K. P. Rodbell, and C. J. Cabral, J. Vac. Sci. Tech. A, **14**(4), 2454-61
11. D. Weiss, O Kraft, and E. Arzt, Mat. Soc. Rec. Proc. **562**, 257-262 (1999).
12. M. J. Kobrinsky, and C. V. Thompson, Acta mater. **48**, 625-633 (2000).
13. R.Spolenak, N.Tamura, B.Valek, A.A.MacDowell, R.S.Celestre, W.L.Brown, J.C.Bravman, H.A.Padmore, T. Marieb, B.W.Batterman' and J.R.Patel, in preparation, presented at MRS Fall Meeting 2000, Boston.

Mat. Res. Soc. Symp. Proc. Vol. 673 © 2001 Materials Research Society

Stress evolution in a Ti/Al(Si,Cu) dual layer during annealing

Ola Bostrom[*][**], Patrice Gergaud[*], Olivier Thomas[*] and Philippe Boivin[**]
[*] Laboratoire TECSEN, UMR CNRS 6122, Université d'Aix-Marseille III
13397 Marseille Cedex 20
[**] STMicroelectronics 6", Rousset, France

ABSTRACT

Mechanical stress and stress evolution in interconnections may cause reliability problems in IC circuits. It is thus of great importance to understand the origin of this stress.
In this paper, the stress evolution during the solid state reaction between blanket titanium and aluminum films has been studied by in-situ substrate curvature measurements. Whereas the formation of $TiAl_3$ is expected to induce large tensile stress because of a global volume decrease of 6-8%, curvature measurements of titanium/aluminum dual layers during annealing at 450°C suggests the formation of a compressive compound.
The evolution of the average force per unit width of the layer during the solid state reaction is interpreted on the basis of a phenomenological model used to describe stress evolution during silicide formation.

INTRODUCTION

The plasticity of thin films has been extensively studied by numerous authors [1,2,3,4] but the effect of a solid state reaction on stress needs further investigation.
In the case of a solid state reaction, such as precipitation inside a matrix, the global volume change when the reactants are replaced by the reaction product may be considered. If the result of a solid state reaction is a global volume decrease, the reaction product is assumed to be in tension.
In this paper, we have studied the effect of the solid state reaction between aluminum and titanium blanket films on the average stress in the thin film system. This system has been chosen because of its importance in the microelectronic industry. The two metals are deposited at different interconnection levels and often a solid state reaction occurs during further processing. This modifies the stress-state in the metal lines and may result in reliability problems.
According to the phase diagram [5], heating of a titanium/aluminum dual layer may result in numerous compounds. However, observations often reveal that $TiAl_3$ is the only phase formed in thin film systems. This may be explained by reaction kinetics [6]. The $TiAl_3$ phase exists in three different structures ($L1_2$, DO_{23} or DO_{22}) including the stable tetragonal DO_{22} observed at high reaction temperatures. The formation of any of the $TiAl_3$ phases results in a global volume decrease of 6-8%. The tensile stress evolution observed by some authors as a result of $TiAl_3$ formation is explained by this volume change [7]. Note however that the systems studied in [7] consisted of a multiple depositions of very thin titanium layers (3at% Ti) between successive aluminum depositions and not two relatively thick ones as in the present study.
In the special case of a reaction between a blanket metal film and a silicon substrate, the silicide formation is often accompanied by a global volume decrease. This should then result in a tensile

silicide but observations often report the opposite [8]. One notes that global volume considerations do not necessarily dictate the stress-state in the reaction product of a solid state reaction. It may be more judicious to take into account the actual volume change at the interface where the reaction occurs. (e.g. Ti is transformed into $TiAl_3$ at the $TiAl_3/Ti$ interface. The resulting volume change is positive).

EXPERIMENTAL DETAILS

A Ti(99.999%) layer (1000Å thick) and an Al(Si 1wt%, Cu 0.5wt%) layer (5500Å thick) were successively sputtered in an ENDURA equipment at C respectively 100°C and 450°C onto 6", 680µm thick silicon wafers covered with a thermal oxide of 8000Å. The titanium was protected from contamination in a base pressure of ~10^{-8} Torr during the 30 seconds preceding the aluminum deposition. After deposition, the thickness of the silicon substrate was reduced to 180µm by mechanical grinding of the backside followed by a chemical etch (10µm) of the damaged zone. Small (10x20mm) samples were then cleaved from these wafers for each of the experiments.
We performed curvature measurements as a function of temperature with a laser reflection arrangement described elsewhere [9]. The force per unit width of layer, induced by the film on the substrate, was deduced from the radius of curvature of the system using the Stoney formula [10]. Different samples were annealed at 450°C during times from 0 to 10 hours in a vacuum of ~10^{-6} Torr. Before and after annealing, each sample was analyzed ex-situ by X-ray diffraction (XRD) on a 4 circles diffractometer ($\lambda_{CuK\alpha}$=1.5418 Å). Both pole figures and symmetric θ-2θ scans were performed. Also asymmetric θ-2θ scans were performed on annealed samples. The tilt (ψ=45°) was chosen to avoid intense reflections from the titanium and the aluminum layers.

RESULTS / DISCUSSION

The titanium and the aluminum both presented a fibre texture. The titanium was preferentially textured (00.2) with some (10.1) grains whereas the aluminum was textured (111). The XRD analysis did not reveal any $TiAl_3$ in the as-deposited sample. After complete consumption of the titanium, pole figures on the aluminum did not show any modification of the texture. An asymmetric scan (ψ=45°) of a fully reacted sample is presented in figure 1. All the observed peaks can be attributed to the stable phase DO_{22} of $TiAl_3$.
In order to follow the evolution of the reaction, we performed ex-situ XRD on samples annealed at 450°C from 0 to 600 minutes. Symmetric and asymmetric θ/2θ–scans permit to conclude that the texture did not evolve during the reaction. The relative intensity of the peaks in the asymmetric scans remained constant.

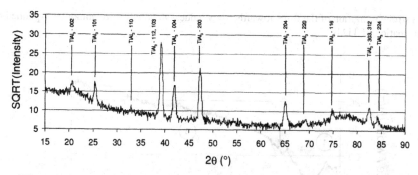

Figure 1 : Asymmetric XRD θ/2θ–scan : all peaks correspond to the DO_{22} phase of $TiAl_3$.

The sum of the four principal (112, 103, 004 and 200) integrated intensities is presented as a function of annealing time in figure 2 where the square of the intensity is plotted in order to put in evidence the diffusion controlled kinetics. The time corresponding to the end of reaction is estimated by extrapolation to approximately 140 minutes.

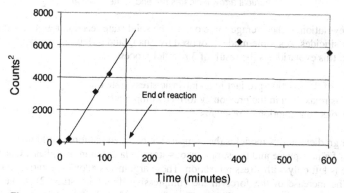

Figure 2 : Intensity of $TiAl_3$ peaks as a function of annealing time at 450°C.

In figure 3, the evolution of the force per unit width is presented as a function of time at 450°C. The sample was heated from RT to 450°C with a rate of 20°C/min resulting in an average force of –80N/m when the temperature of 450°C was attained.

The curve presents 3 distinct regions.
1. During the first 40 minutes, the average force decreases slightly.
2. The force decrease is followed by an increase. The system turns more compressive.
3. A maximum force is reached after 80 minutes. The force then decreases continuously.

The time at which the reaction is finished is indicated by an arrow in figure 3. After the end of the reaction, where the titanium is consumed and the aluminum is completely relaxed, there is still a strong evolution of the force towards a low value. Obviously, this cannot be explained by the formation of TiAl₃.

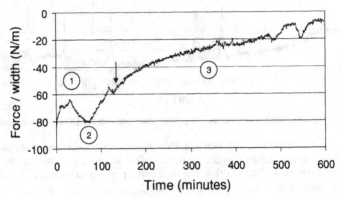

Figure 3 : Evolution of the force per unit width of Ti/Al dual layer during a thermal hold at 450°C. The vertical arrow indicates the end of the reaction.

In fact, the evolution of the average force during the solid state reaction resembles the one often reported for silicides. A simple model proposed by Zhang and d'Heurle [11] explains qualitatively this evolution as the result of 3 parallel processes:

1. Formation of the reaction product in an intrinsic stress state σ_{int}.
2. Continuous relaxation in the previously formed compound.
3. Simultaneous consumption of reactants.

The basic ingredient in this model is the competition between the force build up corresponding to the growth of a new phase and the simultaneous stress relaxation in this phase. Once the reaction is over, one is left only with stress relaxation. These arguments allow a tentative interpretation of figure 3. The increase of the force in the compressive direction (stage 2) is the result of the formation of the compressive TiAl₃. The force added by the compressive compound compensates the elimination of the compressive aluminum. The reaction rate decreases with time resulting in a less and less important addition of compressive stress. During and after the reaction, the relaxation of the compressive stress in the formed compound results in a reduction of the average force in the system.

Assuming stress relaxation by creep, the stress at the time t, in a layer grown at an instant t', may be written :

$$\sigma(t,t') = \sigma_{int} \exp\left(-\frac{(t-t')}{\tau_c}\right)$$

where $t' \in [0,t]$ and τ_c is the relaxation constant of the compound and σ_{int} is the intrinsic stress. The force per unit width of compound is then :

$$dF = \sigma(t,t')dA' \quad \Rightarrow \quad \left(\frac{F}{w}\right)_{compound} = \int_0^h \sigma(t,t')dh' = \sigma_{int}\,\exp\!\left(-\frac{t}{\tau_c}\right)\cdot\left[\int_0^h \exp\!\left(\frac{t'}{\tau_c}\right)dh'\right]$$

where dA' is a surface element on a cross section of the growing compound, w is the width of the film and h' is a length element in the growth direction.

If the growth kinetics is diffusion-controlled the force per unit width of compound may be written :

$$\left(\frac{F}{w}\right)_{compound} = \left[\int_0^h \exp\!\left(\frac{h'^2}{k^2\tau_c}\right)dh'\right]\cdot\sigma_{int}\,\exp\!\left(-\frac{t}{\tau_c}\right)$$

where the time parameter t' has been replaced using : $h'(t)=k\sqrt{t'}$. If one of the elements is consumed, the upper limit of the integral is replaced by the final compound thickness, h_0.

A simulation of the force evolution (neglecting the forces in the two metals) with the constants in table 1, is presented in figure 4.

	h_0 (Å)	k (m*min$^{-0.5}$)	τ_c (min)	σ_{int} (MPa)
TiAl$_3$	3500	3.19*10^{-8}	50 to 150	-500

Table 1 : Constants used in the calculations presented in figure 4.

The point at which the TiAl$_3$ growth stops is well illustrated by the discontinuity of the curves after 150 minutes.

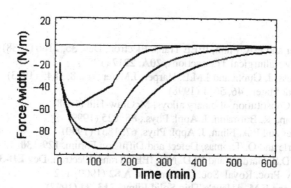

Figure 4 : Evolution of the force per width during the solid state reaction. The smallest relaxation constant, τ_c, corresponds to the least important negative force per width.

The titanium is then completely consumed. The point of maximum tensile or compressive stress is not necessarily at the end of reaction but is determined by the competition between the growth and the relaxation rate. This simple model does not account for the initial evolution in the tensile direction (denoted stage (1) in figure 3). For this first stage of the reaction, a possible explanation is the localised growth of separate $TiAl_3$ grains at nucleation sites as reported in literature. The mechanical behaviour of such buried islands is difficult to determine. In a later stage, when the aluminide layer is continuous, since Al is known to be the most faster diffusing species [12] the new $TiAl_3$ compound layer is thus formed at the $TiAl_3/Ti$ interface, in the titanium layer. The specific volume of the $TiAl_3$ cell is no longer to compare with the volume of three aluminium atoms and one titanium atom but with the volume of only one titanium atom. This volume change is positive. The intrinsic stress in the compound is thus compressive. Finally, this stress is relaxed for longer annealing time.

CONCLUSIONS

We have studied the evolution of the average stress in blanket titanium/aluminum dual layers during annealing at 450°C by curvature measurements and XRD.
Contrary to what is expected from global volume considerations, we believe that the formed $TiAl_3$ is compressive. We propose that the stress evolution of the blanket titanium/aluminum dual layer may be described by the phenomenological model elaborated for stress development during a solid state reaction. The evolution of the average stress during the solid state reaction is the result of four parallel processes :
1. Force evolution in reactants
2. Formation of a compressive compound.
3. Continuous stress relaxation in the formed compound.
4. Simultaneous consumption of reactants.

REFERENCES

[1]   D.S.Gardner and P.A. Flinn, IEEE Trans. Electron. Dev. **35**, 2160 (1988)
[2]   W.D. Nix, Metallurgical Transactions, **20A**, 2217 (1989)
[3]   M.D Thouless, J. Gupta and J.M.E. Harper, J.Mater.Res, **8**, 1845 (1993)
[4]   E. Arzt, Acta Mater., **46**, 5611 (1998)
[5]   M.Hansen, Constitution of binary alloys, McGraw-Hill (1958)
[6]   S. Wohlert and R. Bormann, J. Appl. Phys., **85**, 825 (1999)
[7]   D.S. Gardner and P.A. Flinn, J. Appl. Phys, **67**, 1831 (1990)
[8]   F.M. d'Heurle and O. Thomas, Defect and Diffusion Forum, **129-130**, 137 (1996)
[9]   P.A. Flinn, D.S. Gardner and W.D. Nix, IEEE Trans. Electron. Dev.**ED-34**, 689 (1987)
[10]   G.G. Stoney, Proc. Royal Soc. London Ser. **A 82** (1909) 172
[11]   S.L. Zhang and F.M. d'Heurle,Thin Solid Films, **213**, 34 (1992)
[12]   E.G.Colgan, Materials Science Reports, **5** 1 (1990)

Mat. Res. Soc. Symp. Proc. Vol. 673 © 2001 Materials Research Society

Study of the Yielding and Strain Hardening Behavior of a Copper Thin Film on a Silicon Substrate Using Microbeam Bending

Jeffrey N. Florando and William D. Nix
Dept. of Materials Science and Engineering, Stanford University, Stanford, CA 94305

ABSTRACT

Recently a new microbeam bending technique utilizing triangular beams was introduced. For this geometry, the film on top of the beam deforms uniformly when the beams are deflected, unlike the standard rectangular geometry in which the bending is concentrated at the support. The yielding behavior of the film can be modeled using average stress-strain equations to predict the stress-strain relation for the film while attached to its substrate. This model has also been used to show that the gradient of stress and strain through the thickness of the film, which occurs during beam bending, does not obscure the measurement of the yield stress in our analysis.

Utilizing this technique, the yielding and strain hardening behavior of bare Cu thin films has been investigated. The Cu film was thermally cycled from room temperature to 500 °C, and from room temperature to −196 °C. The film was tested after each cycle. The thermal cycles were performed to examine the effect of thermal processing on the stress-strain behavior of the film.

INTRODUCTION

There is an ongoing need to understand how materials in small dimensions deform, especially in industrial applications. Many current techniques used in industry, such as nanoindentation and wafer curvature, provide some insight into this area. While indentation has proven to be a simple and easy-to-use technique, it fails to provide the fundamental data (yield strength, strain hardening rate) needed for the characterization and modeling of thin film structures. Wafer curvature provides useful information about the elastic and plastic behavior of thin films on substrates, but large temperature changes are required for these experiments. While the thermal cycles experienced during this experiment are similar to the cycles experienced during some chip operations, there is some debate as to how the stress data should be interpreted. One issue is whether the increase in the stress upon cooling is due to strain hardening in the film [1], or due to the temperature dependence of the yield stress [2-3].

An improved microbeam bending technique may be able to help answer this question. The method is similar to previous work done on microbeam bending [4-5], except that triangular-shaped silicon microbeams are used. These triangular beams have the advantage that the entire film on the top surface of the beams is subjected to a uniform state of strain when the beams are deflected, unlike the standard rectangular geometry where the bending is concentrated at the support. The uniform strain state allows for the calculation of the stress-strain behavior of the film. Since the test is performed at room temperature, the film's yielding behavior can be examined independent of temperature, leading to a possible method of determining the mechanism that causes the increase in stress in the wafer curvature experiments. The first step in this process is to determine if this testing method has the sensitivity to detect changes in the stress-strain behavior of a film with different initial stress states.

THEORY

The advantage of the triangular beam is the constant moment per unit width acting in the beam during deflection. This constant moment per unit width results in a constant curvature along the beam. Therefore, a film will yield at all points along the beam at the same time, unlike rectangular beams where the film yields first at the support.

Although the triangular geometry eliminates the strain variation along the length of the beam, there is a linear variation in strain through the thickness of the film. Since the substrate is four times as thick as the film, the approximation can be used that there is no variation in stress or strain through the thickness of the film, and the film can be treated as if it were in uniform tension. Using this approximation, we can derive the following average stress and strain equations:

$$\overline{\delta\sigma} = \frac{L\delta P}{w_o \int\limits_{ts}^{ts+tf}(y-y_o)dy} \tag{1}$$

$$\overline{\delta\varepsilon}_{xx} = \frac{2\delta u}{t_f L^2} \int\limits_{ts}^{ts+tf}(y-y_o)dy \tag{2}$$

where δP is the incremental load associated with deformation of the film (where the load increments needed to deflect the substrate have been subtracted), δu is the incremental displacement, L is the length of the beam, w_o is the base width, t_f is the film thickness, t_s is the substrate thickness, and y_o is the position of the neutral plane for bending. All of these values are measured quantities except for the neutral plane positions, which are calculated. To understand why the neutral plane position changes, we use the following thought experiment. For a film on a substrate, the location of the neutral plane for bending will be based on a ratio of the substrate and film thickness and their respective moduli. If the bi-layer is deformed elastically the neutral plane will not move. When the film begins to yield, however, it supports a lower percentage of the load, and the neutral plane moves towards the middle of the linear elastic substrate. The neutral plane positions are calculated for the beam-bending problem by dividing the film into many horizontal segments and assuming a Ramburg-Osgood stress-strain law. The average stress and strain approximations are not used for the calculation of the neutral plane positions. A more detailed description of the methodology for this calculation is described elsewhere [6].

X-ray strain measurements were used to determine the residual stress of a film on a substrate. For a textured film, $\sin^2\psi$ measurements can be used to find the strain in the film [7]. The strain for a film with a dual texture can be found by performing $\sin^2\psi$ measurements on a second set of planes. The d-spacings from different families of planes can be compared by using the normalized quantity $d_{hkl}\sqrt{h^2+k^2+l^2}$ [8]. This normalized quantity should vary linearly with $\sin^2\psi$ for a particular texture.

EXPERIMENTAL PROCEDURE

The beams were deflected using a commercial Nano II nanoindenter from the MTS Nano Innovation Center, which accurately measures the load and displacement of the beams. The experimental data is used to calculate the stress-strain behavior of the film by following the methodology described in the previous section.

A Philips diffractometer was used to measure the residual stresses in the film. Since the Cu film was deposited on a thermal oxide layer, it has a dual <111>, <100> texture. The biaxial elastic strain in the <111> textured grains was measured by measuring the <111> d-spacing for $\psi = 0°$ and then rotating sample through the angle $\psi = 70.5°$, and re-measuring the <111> d-spacing. For the <100> textured grains, the <100> d-spacing was measured at $\psi = 0°$, and then the sample was rotated to $\psi = 50.7°$ and the <111> reflection was measured. The d-spacings were normalized to obtain the biaxial elastic strain. The measured strains were converted to average stresses by assuming an average biaxial modulus.

After the film was tested at room temperature, the film was thermally cycled to 500° C in vacuum, with a base pressure of 1.5×10^{-5} torr. The residual stress was measured and the beams were tested again. In order to achieve a different initial deformation state, the film was cooled with a two-minute dip in a liquid nitrogen bath, and then heated back to room temperature. The residual stress was again measured, and the film was tested using the beam bending method.

RESULTS AND DISCUSSION

The premise of this technique relies on the ability to extract the stress-strain behavior of the film. Therefore, the approximation used to derive the average stress-strain equations must be verified. To do this, we choose a fictitious film that obeys a linear hardening stress-strain law. This law is chosen because it has an abrupt yield point. Since the film's stress-strain behavior is known, then its theoretical load-displacement behavior can be calculated. This calculated "data" is then inserted into the average stress-strain equations (eqns. 1-2). Figure 1 shows a plot of the linear hardening law in comparison with the average stress-strain equations. The graphs should be identical, and as seen in Figure 1, the curves match very well. There is a slight rounding in the average equation curve at the yield point, but that can be expected since the stresses and strains are being averaged. For this case, using the average stress-strain equations is a valid approximation.

The film was mechanically cycled to examine the effects of the initial stress state on the ensuing stress-strain behavior. Figure 2 shows a film that has been mechanically cycled three times. The stresses and strains applied are similar to those experienced during a thermal cycle. The first cycle shows elastic loading, followed by yielding that can be characterized by two regions. The initial region shows a high work hardening rate, which is followed by a nearly flat curve and a low rate of work hardening. There is a hold segment at the end of the loading cycle that shows room temperature creep. On unloading, the film yields in compression even though the overall film stress is still tension. This Bauschinger effect is probably associated with the strong polarity of the deformation microstructure in the film. Strong Bauschinger effects have been reported in studies of plasticity in thin metal films on substrates [3]. The 2nd loading cycle begins while the film is at a compressive stress of about 150 MPa. The overall shapes of the 2nd and 3rd cycles are similar, with only a small difference in the initial and final stresses. These cycles are markedly different from the 1st cycle, and this difference can be attributed to the initial deformation state. Since the 2nd and 3rd cycles started in residual compression instead of tension, the film must first unload elastically in compression and then load elastically in tension before

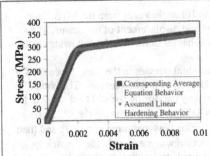

Figure 1- Comparison of average stress-strain equations to the theoretical linear hardening model.

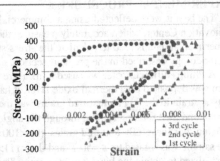

Figure 2- Cyclic loading stress-strain plot for a 1μm thick Cu film.

yielding. Therefore, there is a larger elastic region in the 2nd and 3rd cycles than in the 1st cycle. The difference in the stress-strain behavior between the cycles exhibits the sensitivity that this technique has to detect the effect of changes in the deformation microstructure on the behavior of the film.

A summary of the residual stress measurements by the x-ray strain technique is given in Table I. The x-ray data shows that the strains in the <111> and <100> textured grains are equal, which verifies the assumption that the same misfit strain occurs in all the grains. Since the strains are equal, the stresses are different in the different oriented grains. In this study however, we are interested in the average properties of the film and therefore an average modulus was used to calculate the average residual stress in the film.

The x-ray measurements show that initially the film had a tensile residual stress after the deposition process. After annealing the sample and cooling back to room temperature, the film had an increase in the tensile residual stress due to grain growth, densification, and plastic deformation that occurred at elevated temperatures. Upon cooling the sample to liquid nitrogen temperature and heating back to room temperature, the film had a residual compressive stress. This can be explained by assuming that during the cooling cycle, the film continued to yield in tension. Upon re-heating, the film elastically unloaded the tension stresses and then began to deform in compression. We should note that the x-ray strain technique measures the residual stress in the film when it is supported by the massive substrate. In our beam bending experiments, the Si substrate is 4.2 μm, while the film is 1.0 μm. Since the substrate and film are of similar thickness, the substrate will now accommodate some of the misfit strain. The misfit strain is mainly due to the differences in thermal expansion coefficients between Si and Cu. For the case of the Cu film on top of the massive substrate, the substrate is so thick that the Cu film

Table I. Residual Stress Measurements

Temperature Cycle	Residual Stress
As-Deposited	+170 MPa +/- 4 MPa
RT -> 500 °C -> RT	+260 MPa +/- 16 MPa
RT ->-196 °C -> RT	-30 MPa +/- 5 MPa

Table II. Calculated Residual Stress Measurements on the Beam

Cycle	Residual Film Stress on Wafer	Calculated Residual Film Stress on Beam
As-Deposited	170 MPa	80 MPa
500 °C to RT	260 MPa	120 MPa

will accommodate most of that strain. In the case of a Cu film on the Si beam, the Si substrate can accommodate some of the strain, and consequently, the stress in the film will be reduced. For the as-deposited film and the film thermally cycled to 500 °C, the assumption is made that the same misfit strain occurs regardless of the whether the film is on the massive substrate or on the beam. For this case, the average residual stress on the beam can be calculated, and the results of this calculation are shown in Table II. The residual stress in the film after the liquid nitrogen cool could not be calculated because the same misfit strain assumption cannot be made, and the misfit strain that occurs on the beam during the cooling process is not known.

Knowing the residual stresses permits a measurement of the actual stress and strain in the film. The stress-strain curve for the as-deposited film, and for the film that was annealed to 500 °C is shown in Figure 3. The annealed film, with a higher residual stress, yields at a lower overall stress (0.2% offset yield stress of 355 MPa) than the as-deposited film (0.2% offset yield stress of 400 MPa), probably due to the larger grain size. Focus ion beam (FIB) microscopy shows that the grain size of the as-deposited film was 0.7 μm, while the grain size for the annealed film was 0.96 μm.

A comparison of the liquid nitrogen cooled sample and the annealed sample is shown in Figure 4. Since the residual stress on the beam after the liquid nitrogen cool could not be calculated, the plot is given as Δ (stress) versus strain and is a relative measure of the stress-strain behavior. The data shows that the behaviors are very similar. Table I, however, shows that the liquid nitrogen sample is expected to be in residual compression, while the annealed sample is in residual tension. If this were the case, then the liquid nitrogen curve should show a larger region of elastic loading before it yields. Since the two curves look similar, this suggests that they have a similar residual stress state. Therefore, the film on the beams after the liquid nitrogen

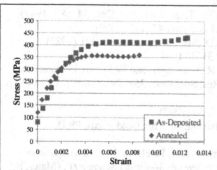

Figure 3- Stress-strain plots of 1 μm Cu film comparing as-deposited vs. annealed.

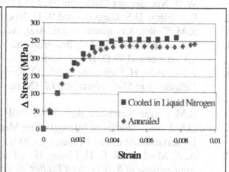

Figure 4- Δ stress-strain plots for 1 μm Cu film comparing annealed vs. liquid nitrogen cooled.

cool is still in residual tension, even though the film far away from the beams, on the massive substrate region, is in residual compression. Optical microscopy also confirms that the beams are in residual tension since the beams are curved upward. A possible explanation for this difference is that since the beam substrate accommodates some of the misfit strain, then during the cooling cycle, the beam does not yield and thus elastically loads and unloads to the same residual stress state. The residual stress on the beams needs to be measured using a synchrotron radiation x-ray technique [9].

CONCLUSIONS
The average stress-strain approximation used to determine stress-strain behavior of a film on a substrate is shown to be valid. Using that methodology the microbeam bending technique was used to study the effect of the initial deformation state on the stress-strain behavior of the film. This study showed that by mechanically cycling the film, the stress-strain behavior changes as a function of the initial deformation state and this method has the sensitivity to detect these differences. The initial deformation states were also changed by thermally cycling the film, and differences between the as-deposited and annealed state were observed. When cooled in liquid nitrogen, however, the film on the beams appear to remain in residual tension and no observable difference in behavior was noticed between the annealed and the cooled sample.

ACKNOWLEDGMENTS
The authors would like to thank the Intel Graduate Fellowship Program and the Department of Energy (DE-FG03-89ER45387) for funding this project. A special thanks goes to Qing Ma, Harry Fujimoto, and the R1 staff in Components Research at Intel Corp. for help in processing of the wafers, to Oliver Kraft and Ruth Schwaiger from the Max-Planck-Institut für Metallforschung in Stuttgart, Germany for their useful discussion and FIB pictures, and to Bryan Tracy and Jonnie Barragan from Advanced Micro Devices (AMD) for the use of their FIB machine. This work made use of the National Nanofabrication Users Network facilities funded by the National Science Foundation under award number ECS-9731294.

REFERENCES
1. W.D. Nix, Scripta Materialia **39**, 545 (1998).
2. P.A. Flinn, D.S. Gardner, and W.D. Nix, IEEE Trans. on Electr. Dev. **34**, 689 (1987).
3. R.M. Keller, S.P. Baker, and E. Arzt, Acta Materialia **47**, 415 (1999).
4. T.P. Weihs, S. Hong, J.C. Bravman and W.D. Nix, J. Mater. Res. **3**, 931 (1988).
5. S.P. Baker and W.D. Nix, J. Mater. Res. **9**, 3131 (1994).
6. J. Florando, H. Fujimoto, Q. Ma, O. Kraft, R. Schwaiger, and W.D. Nix in *Materials Reliability in Microelectronic IX*, (Mater. Res. Soc. Proc. **353**, Pittsburgh, PA, 2000) pp. 231-236.
7. B.M. Clemens and J.A. Bain, MRS Bull. 17, 46 (1992).
8. O.S. Leung, Studies of Strengthening Mechanisms in Thin Gold Films, Ph.D. Dissertation, Stanford University, (2001).
9. A. A. MacDowell, C. H. Chang, H.A. Padmore, J. R. Patel, and A. C. Thompson in *Applications of Synchrotron Radiation Techniques to Materials Science IV*, (Mater. Res. Soc. Proc. **524,** Pittsburgh, PA, 1998) pp. 55-58.

Mat. Res. Soc. Symp. Proc. Vol. 673 © 2001 Materials Research Society

Mechanical behavior of thin Cu films studied by a four-point bending technique

Volker Weihnacht and Winfried Brückner
Institute for Solid State and Materials Research Dresden, Helmholtzstrasse 20, D-01069
Dresden/ Germany, v.weihnacht@ifw-dresden.de

ABSTRACT

Four-point bending experiments in combination with thermal cycling of thin films on substrates were performed in a dedicated apparatus. Strains up to ±0.8% could be imposed into Cu films of 0.2, 0.5, and 1.0 μm thickness on Si substrates by bending the substrates at various temperatures in high vacuum. After relief of the bending, the residual stress was measured by the wafer-curvature method. At temperatures below 250°C, the yield behavior is asymmetric in tension and compression. The amount of plastic strain introduced by external bending increases with film thickness, but the absolute values of the introduced plastic strains are very low throughout. At higher temperatures, there is no clear thickness dependence and no asymmetry in tension and compression. The results are discussed in connection with the formation of misfit dislocations during plastic deformation of thin films.

INTRODUCTION

Thin metallic films play an important role in many modern technologies due to their electrical, magnetical and optical properties. The ongoing miniaturization of the used structures leads to stability problems due to cracking, hillock and void formation, and local plastic deformation. Therefore, the mechanical properties of thin metallic films have been intensively studied and modeled.

For mechanical testing of thin films some methods from bulk testing have been adapted to the thin film geometry, e.g. micro-tensile testing [1] and nanoindentation measurement [2]. Also new methods were developed, such as bulge testing [3], micro-beam bending [4], tensile testing of films on flexible or plastic substrates [5], and stress measurement during thermal cycling [6]. The most common technique is thermal cycling of films on Si substrates. It gives a good insight into the plastic behavior under thermal stresses, but is hampered by the fact that temperature and film stress can not be varied independently.

It turned out that thin-film plasticity is different from bulk plasticity for the same material. In particular, the difference shows up in very high film strengths. This could not be explained by microstructural reasons, because also pure single-crystalline films are significantly stronger than the corresponding bulk material [7]. Further peculiarities of mechanical thin-film behavior are effective strengthening during cooling and a Bauschinger effect with low compressive stresses during reheating in the thermal cycling experiment [8].

In this work, in addition to the thermal cycling behavior, information was obtained by imposing external stresses into Cu films on Si samples by four-point bending. This technique enabled us to measure the isothermal stress-strain behavior at various temperatures. The results are compared with stress measurement during thermal cycling between room temperature and 500°C. Both the thickness dependence of yield stress and other features of the stress curves are discussed with view on thin-film dislocation models.

EXPERIMENT

Si wafers with a thickness of 380 µm, oxidized and polished on both sides, were used as substrates. The wafers were cut into strips with 60×7 mm² in size. The strips were mechanically treated in order to improve the resistance against cracking during four-point bending. This treatment consisted of grinding and polishing the side faces of each strip. Blanket Cu films of 0.20, 0.53, and 1,03 µm thickness were deposited on the Si-strips by magnetron sputtering ($2×10^{-7}$ mbar base pressure, $2×10^{-3}$ mbar working pressure, and 4 kW sputter power). By using a mask, only the central area of 30×7 mm², where the bending stress is constant, was coated.

The film stress was determined by the wafer-curvature technique (WCT) in a dedicated in-house designed apparatus, described in Ref. 9. The stress values were obtained from Stoney's equation [10] by measuring the substrate curvature with a laser-optical system. The annealing and thermal cycling was performed in high vacuum ($<5×10^{-5}$ mbar) between room temperature (RT) and 500°C with a rate of 4 K/min. During cooling below 150°C, this rate decreased considerably due to the heat capacity of the furnace. During wafer-curvature-stress measurement, the samples were placed on the lower bearings of the four-point bending equipment.

Four-point bending was performed by shifting the lower bearings in height. The lateral spacing between the upper bearings s_1 amounted to 30 mm, that between the lower and upper ones s_2 10 mm. The accuracy of the height shift h of the lower bearings was better than ±0.01 mm to allow for precise bending.

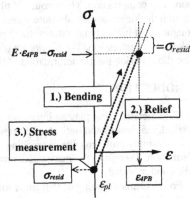

The bending experiments consisted of repeated bending-relief cycles with stepwise increased bending load. After each bending-relief cycle, the residual stress σ_{resid} was measured by the WCT using Stoney's equation revised for uniaxial stress. In the following cycle, the shift was increased by Δh=0.25 mm. This procedure was successively continued up to h=8.0 mm.

The strain ε_{4PB} in the film on the substrate can be calculated from the distances s_1 and s_2, the shift h, and the thickness of the Si substrates, t_s, by the equation

Figure 1. Schematic illustration of the stress-strain curves of the film during one bending-relief cycle of the four-point bending experiment.

$$\varepsilon_{4PB} = \frac{t_s}{s_1 s_2 + \frac{2}{3} s_2^2} h .$$
(1)

Assuming that the unloading curve has linear elastic behavior, the film stress σ_{4PB} at the maximum bending strain ε_{4PB} can be estimated as

$$\sigma_{4PB} = E \cdot \varepsilon_{4PB} - \sigma_{resid} ,$$
(2)

where E designates the Young's modulus of the film.

Prior to mechanical testing, the films were annealed in a thermal cycle up to 500 °C in order to stabilize the microstructure. After that, the microstructure was investigated by transmission electron microscopy (TEM) and focussed ion beam microscopy (FIB). Texture measurements were carried out by x-ray diffraction (XRD).

RESULTS AND DISCUSSION

After vacuum annealing of the films in a cycle to 500°C, log-normal grain-size distributions with median grain sizes of the order of the film thickness were found. All films had pure {111} fiber textures with a full width at half maximum of about 11°. In the TEM analysis, parallel dislocation arrangements and dislocation entanglements were observed in some grains.

After annealing in a thermal cycle up to 500°C, the stress was measured during a second thermal cycle. The stress vs. temperature curves displayed all typical features of thermal cycling curves: (i) thermoelastic behavior in the beginning of heating, (ii) early deviation from the thermoelastic line (Bauschinger effect), (iii) compressive stress plateau up to high temperatures, and (iv) thickness-dependent strengthening during cooling. The strengthening is most pronounced in the thinnest film. After completion of the thermal cycle, the film stresses reached nearly their initial values. Thermal cycling curves and discussion of related deformation mechanisms are reported in many papers (e.g. [6,11]).

It has to be considered that four-point bending causes a uniaxial stress, in contrast to the biaxial thermal film stress. In order to avoid a superposition of uniaxial bending stress and biaxial thermal stress, the bending experiments were carried out at states free of thermal stress. This was obtained by convenient thermal pretreatments. The simplest way to a stress-free state is to heat the sample from the initial tensile stress to about 200°C (see Fig. 2). In order to obtain stress-free states at different temperatures, different thermal pretreatments have to be used. The thermal pretreatments used in our case are illustrated in Fig. 2 for the example of the 0.5 µm thick film. The first one consisted of cooling to –196°C in liquid nitrogen and heating to that temperature where the film stress was zero (100°C in the 0.5 µm film), and the other one of heating the sample up to about 370°C and cooling to 340°C. In table I, the temperatures of stress-free states after the pretreatments of all films are summarized. After the special thermal pretraetment for each film, isothermal bending experiments were carried out as described above and illustrated in Fig. 1. The results of the bending experiments are represented in Fig. 3. The curves show the residual stress after each bending-relief cycle versus the stepwise increased bending strain. From these residual stresses, the stress-strain curves in Fig. 4 were calculated using Eq. 2.

	1.0 µm	0.5 µm	0.2 µm
After cooling to –196°C and reheating (1)	20°C	100°C	200°C
After heating from room temperature (2)	170°C	210°C	250°C
After heating to 370°C and cooling to 340°C (3)	340°C	340°C	340°C

Table I. Temperatures, at which stress-free states were obtained after thermal pretreatment (see also Fig. 2). At these temperatures, the four-point bending experiments were carried out.

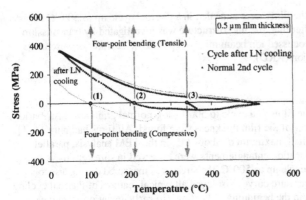

Figure 2. Stress-temperature curves of the 0.5 µm thick Cu film during the thermal pretreatments. By three different pretreatments, stress-free states were achieved at three different temperatures: (1) cooling in liquid nitrogen (LN), (2) heating from room temperature, and (3) interrupting heating at about 370°C and cooling to 340°C.

There is especially one remarkable feature in the σ_{resid} vs. ε_{4PB} curves at relatively low temperatures (see Fig. 3a and 3b): an asymmetry of the residual stresses in tension and compression. Compressive bending leaves residual stress beginning at small bending strains. Tensile bending, on the other hand, does not leave significant residual stress in the beginning. Only after exceeding a "critical" tensile bending strain of about 0.3%, significant residual stresses are left. This phenomenon applies to all three film thicknesses, but is most pronounced in the thinnest film. Once a residual stress is left, σ_{resid} increases almost linearly with ε_{4PB}. The amount of σ_{resid} is larger for increasing film thickness. Both features, the asymmetry and the thickness dependence of σ_{resid}, apparently vanish at high temperatures (see Fig. 3c).

The magnitude of σ_{resid} represents the amount of plastic deformation the film has undergone during bending. Hence, the increase of σ_{resid} with increasing film thickness indicates a decrease of flow stress with increasing film thickness. This dependence corresponds qualitatively to the frequently observed inverse proportionality of flow stress and film thickness, that is commonly account by the dimensional constraint of dislocation plasticity in thin films (e.g. [12,13]).

To account for the asymmetry of residual stresses, we propose an explanation based on the misfit-dislocation model in thin films. Misfit dislocations are accumulated at the film-substrate interface during plastic deformation, because dislocations cannot penetrate into the substrate. The density of misfit dislocations is directly related to a certain amount of plastic strain ε_{pl} [14,15]. Misfit dislocations form during cooling in the thermal annealing cycle (see point A in Fig. 5). The accumulation of misfit dislocations is connected with a strengthening due to the interaction of dislocations with misfit dislocations [14,15]. Reheating of the film does not change the dislocation configuration (see points C and D) in the beginning. Further heating leads to a Bauschinger effect, since "easy" plastic deformation is yielded by back gliding of the accumulated misfit dislocations (see point E). In the four-point bending experiment, the dislocation configuration of point D is present at the beginning (see Fig. 6). Bending in tensile direction does not cause plastic deformation at first, because of the repulsive interaction with the misfit dislocations. Only at large bending strains, the interaction can be overcome (see point C) and plastic deformation occurs by depositing further misfit dislocations into the dislocation arrangement at the interface. Compressive bending, on the other hand, is immediately connected with plastic deformation, because the plastic strain is yielded by the back glide of misfit dislocations. This effect is, therefore, equivalent to the Bauschinger effect during thermal cycling [8]. No clear asymmetry of the stress can be observed at 340°C (see Fig. 3c). At such high temperatures, dislocation mediated plasticity is probably not the decisive deformation mechanism. Diffusional creep is rather assumed to be responsible for the stress relaxation. There is no indication for a difference of diffusional creep in tension and compression.

A further question worth to be discussed is the small deviation of the calculated stress-strain curves from the elastic line up to large bending strains (see Fig. 4). The calculated σ_{4PB} values are in obvious contradiction to the other results. Especially the extremely large stresses up to 800 MPa were not nearly found during thermal cycling. There is only one explanation for this contradiction: the unloading was not elastic, and, hence, the calculated σ_{4PB} values are strongly overestimated.

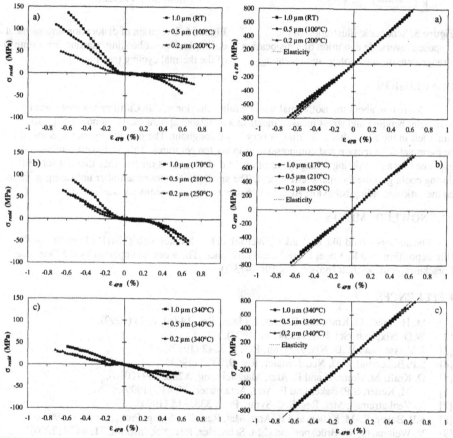

Figure 3. Residual stress vs. four-point-bending strain with stepwise increased strain for 0.2, 0.5, and 1.0 μm film thickness after different thermal pretreatments: a) cooling in liquid nitrogen, b) heating from RT, and c) interrupting heating at about 370°C and cooling to 340°C.

Figure 4. From the results of Fig. 3 calculated stress-strain curves. The σ_{4PB} values were calculated by Eq. 2 assuming linear-elastic unloading behavior.

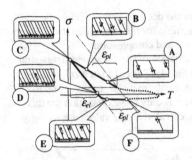

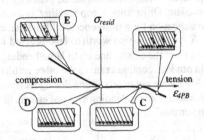

Figure 5. Schematic illustration of the supposed reversible evolution of a dislocation arrangement during cooling and reheating.

Figure 6. Evolution of dislocation arrangement during four-point bending, starting from state D of the thermal cycling (see Fig. 5).

CONCLUSION

Information about the isothermal stress-strain behavior of thin Cu films was obtained by four-point bending and stress measurements in a dedicated apparatus. The amount of plastic strain left in the films after unloading is very low throughout. The plastic behavior was found to be asymmetric in tension and compression at lower temperatures. This asymmetry might be explained by a preexisting arrangement of misfit dislocations at the interface that has formed during cooling in the thermal pretreatment. The small plastic strains left after unloading point to an inelastic unloading behavior. At higher temperatures, no asymmetry was found.

ACKNOWLEDGMENTS

The authors would like to thank C. Wenzel and U. Merkel with TU-IHM Dresden for the film deposition, and R. Vogel for technical assistance. The work is supported by the Deutsche Forschungsgemeinschaft (project No. BR 1473/2).

REFERENCES

[1]   M. Hommel, O. Kraft, and E. Arzt, J. Mater. Res. **14** (6), 2373 (1999).
[2]   W.D. Nix, Mat. Sci. Eng. **A234**, 38 (1997).
[3]   J. Vlassek and W.D. Nix, J. Mater. Res. **7**, 3242 (1992).
[4]   S.P. Baker and W.D. Nix, J. Mater. Res. **9**, 3131 (1994).
[5]   O. Kraft, M. Hommel and E. Arzt, Mat. Sci. Eng. **A288**, 209 (2000).
[6]   R.-M. Keller, S.P. Baker und E. Arzt, Acta mater. **47**, 415 (1999).
[7]   R. Venkatraman, Mat. Res. Soc. Symp. Proc. **338**, 215 (1994).
[8]   S.P. Baker, R.-M. Keller, and E. Arzt, Mat. Res. Soc. Symp. Proc. **505**, 605 (1998).
[9]   V. Weihnacht, W. Brückner and C.M. Schneider, Rev. Sci. Instrum. **71**, 4479 (2000).
[10]   G. Stoney, Proc. R. Soc. London, **A82**, 172 (1909).
[11]   R.P. Vinci, E.M. Zielinsky und J.C. Bravman, Thin Solid Films **262**, 142 (1995).
[12]   W.D. Nix, Metall. Trans. **20A**, 2217 (1989).
[13]   C.V. Thompson, J. Mater. Res. **8** (2), 237 (1993).
[14]   W.D. Nix, Scripta mater. **39**, 545 (1998).
[15]   V.Weihnacht and W. Brückner, to be published in Acta Mater..

Discrete Dislocations:
Observations and Simulations

Mat. Res. Soc. Symp. Proc. Vol. 673 © 2001 Materials Research Society

DISLOCATION DYNAMICS SIMULATIONS OF DISLOCATION INTERACTIONS IN THIN FCC METAL FILMS

PRITA PANT*, K.W. SCHWARZ**, S.P. BAKER*
*Materials Science & Engineering Department, Cornell University, Ithaca, NY 14853.
** IBM Research, Yorktown Heights, NY, 10598.

ABSTRACT
Mesoscopic simulations of dislocation interactions in thin, single crystal FCC metal films were carried out. Interactions between threading-misfit and threading-threading dislocation pairs were studied and the strength of the interactions determined. Threading-threading interactions were found to be significantly stronger than threading-misfit interactions. Dislocations with different possible combinations of Burgers vectors were studied under cyclic loading. Only annihilation of dislocations was seen to result in residual dislocation structure after complete unloading. No differences were observed in the nature of threading-misfit interactions in 111 and 001 oriented films.

INTRODUCTION
The mechanical properties of metals at small scales have been a topic of increasing interest as sizes of devices have been shrinking. A well known example is that of metallizations in integrated circuits. Common reasons for the failure of metallizations are stress controlled processes like decohesion, cracking, void formation and electromigration. It would be useful then, to be able to estimate stresses and, based on mechanisms of failure, predict the strength and/or lifetime of metallizations under given working conditions. But metal thin films have different dimensional and microstructural constraints than bulk metals and show quite different mechanical properties [1].

A standard experiment for investigating mechanical properties involves thermally cycling a thin metal film on a substrate. Stresses arise due to differential thermal expansion. Fig. 1 shows the stresses measured in a 1μm thick, passivated Cu film during a thermal cycle. Similar behavior has been observed in other studies [2, 3]. Compared to bulk Cu, thin films support higher stresses at both low and high temperatures and show higher strain hardening rates on cooling. Also, thin films exhibit "negative yielding" *i.e.* the film yields plastically (deviates from the thermoelastic line) in compression, while the stress is still tensile.

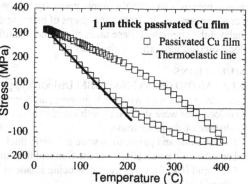

Fig 1. Stresses measured in a passivated Cu film during thermal cycling [9].

Nix [1] and Freund [4] have proposed models that explain the high strength of thin films based on an energy argument. According to these models, a threading dislocation moving on its glide plane in a passivated film will deposit misfit segments at both interfaces (Fig. 2). The dislocation will move into the film when the critical or "channeling" stress (σ_{ch}) is reached. This stress is determined by the point at which the strain energy relieved by the dislocation is just equal to the energy needed to create the misfit dislocations, and is given by

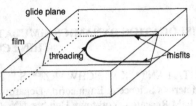

Fig 2. Nomenclature for dislocations

$$\sigma_{ch} = A\mu_{eff}\frac{1}{h} , \qquad (1)$$

where A is a geometric factor, μ_{eff} is the effective shear modulus and h is thickness of the film.

Baker *et al.* have used this energy argument to explain "negative yielding" [5]. For stresses below σ_{ch}, it is energetically infeasible for the dislocations to stay in the film. Dislocations driven into the film during loading should run back out of the film (compressive plastic strains) if the stress is reduced to levels less than σ_{ch} even if the stress is still tensile.

Other models consider that blocking of threading dislocations by misfit dislocations lying on intersecting glide planes contributes to the high stress levels [6] and strain hardening rates [7] of thin films. In these models, analytical methods based on straight dislocations are used to estimate the strength of threading-misfit dislocation interactions. Such methods, however, do not take the actual shape of the dislocations into account. In a real film, dislocations will move around on their glide planes, adopting complex curved shapes in order to reduce the total energy in the film as they interact. With advances in computational ability it has been possible to study these interactions in detail using simulations. Fully three-dimensional simulations provide us with a tool to understand dislocation mechanisms and their connection to thin film plasticity.

Our aim is to study dislocation-mediated plastic deformation in annealed thin FCC metal films using discrete Dislocations Dynamics (DD) simulations. Such films typically have a columnar grain structure [8, 9], so the problem of dislocation interactions becomes quasi two-dimensional. Also, 111 and 001 are the most common grain orientations for FCC films [10, 11]. This provides us with suitably simple geometry for the dislocation interaction problem. In this paper we present results from simulations of pairwise dislocation interactions. We have calculated the strength of these interactions and identified some interactions resulting in residual dislocation structures after cyclic loading.

SIMULATIONS

The PARANOID (Parallel Nodal IBM Dislocation) program [12] was used for the simulations. Simulations were set up with the following parameters and boundary conditions:

- Dislocations were stopped at both interfaces, *i.e.* were not allowed to penetrate into the substrate or the passivation.
- The substrate and passivation were infinitely thick.
- The substrate, passivation, and the film had the same elastic constants.
- The metal (Cu) film was treated as being elastically isotropic with Young's modulus $E = 110$ GPa and Poisson's ratio $\nu = 0.3$.
- The film was 800 nm thick.
- Equal biaxial strain was applied in the plane of the film.
- Dislocations were only allowed to glide on their slip planes.

An FCC metal film with a {111} plane parallel to the film plane (referred to as a 111 film in this paper) has three inclined {111} glide planes, as shown in Fig. 3. Similarly a film with an {001} plane parallel to the plane of the film (an 001 film) has four inclined {111} glide planes.

The possible interactions between pairs of dislocations in 111 or 001 films, can be classified into three broad categories:

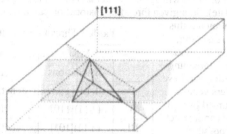

Fig 3. Glide planes in 111 film form a pyramid. Two intersecting glide planes are shaded

1. Threading-Misfit interactions: A threading dislocation moving through the film may interact with misfit dislocations deposited by a threading dislocation that previously ran by on an intersecting glide plane (Fig. 4).

2. Threading-Threading interactions: Threading dislocations could also interact with other threading dislocations (Fig. 5)

3. Interactions between threading dislocations and dislocations with Burgers vectors in the plane of the film: Dislocations with Burgers vectors in the plane of the film experience zero resolved shear stress in an in-plane, biaxial stress state and hence do not move. These interactions have not been considered in this paper.

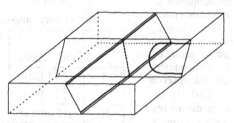

Fig 4. Threading-misfit interaction

Simulations of threading-misfit and threading-threading interactions are reported here. In both cases, the interacting dislocations could be on parallel (*e.g.* Fig. 5) or intersecting (*e.g.* Fig. 4) glide planes. The glide planes and Burgers vectors of the interacting dislocations were specified, a fixed

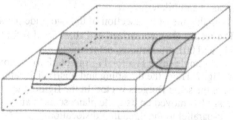

Fig 5. Threading- threading interaction

strain was applied and dislocation interactions observed. The strength of the interaction (σ_{int}) is defined as the stress required to just drive the threading past the misfit (or the threading dislocations past each other). The film was then unloaded and interactions resulting in some residual dislocation structure identified.

RESULTS

Threading-misfit interactions

In both 111 and 001 oriented films, for a given threading dislocation (glide plane and Burgers vector) there are 18 possible interactions with misfit dislocations on the remaining 3 glide planes (3 glide planes, each with 6 Burgers vectors considering positive and negative Burgers vectors as two distinct cases). 10 of these misfit dislocations (6 dislocations in the plane of the film and 4 on inclined planes) have Burgers vectors in the plane of the film and are not considered here. The remaining 8 dislocations can be classified into 4 pairs of positive and negative Burgers vectors.

Only 4 of these will "grow" under an applied tensile stress (the other 4 will "shrink" and annihilate). So a given threading dislocation can interact with 4 distinct misfit dislocations.

Following this line of reasoning, for any given glide plane the threading dislocation can have 2 Burgers vectors. Each threading dislocation can interact with 4 possible misfits resulting in 8 distinct cases of threading-misfit dislocation interactions. These 8 cases were simulated and the nature of the dislocation interactions was observed.

Dislocations either did not react, reacted to form a junction, or reacted so as to annihilate along the line of intersection of the two glide planes. The results are summarized in Tables I and II for 111 and 001 orientations, respectively.

Table I Threading-misfit interactions for 111 oriented film

Case	Misfit dislocations	Threading dislocation $(1\bar{1}1)$ [011]	Threading dislocation $(1\bar{1}1)$ [110]
I	$(\bar{1}\bar{1}1)[\bar{1}0\bar{1}]$	Junction formation	Junction formation
II	$(\bar{1}11)[110]$	No reaction	Annihilation
III	$(\bar{1}11)[101]$	Junction formation	Junction formation
IV	$(1\bar{1}1)[011]$	Annihilation	No reaction

Table II Threading-misfit interactions for 001 oriented film

Case	Misfit dislocations	Threading dislocation $(1\bar{1}1)$ [011]	Threading dislocation $(1\bar{1}1)[\bar{1}01]$
I	$(\bar{1}\bar{1}1)[101]$	Junction formation	No reaction
II	$(\bar{1}\bar{1}1)[011]$	Annihilation	Junction formation
III	$(111)[0\bar{1}1]$	No reaction	Junction formation
IV	$(111)[\bar{1}01]$	No reaction	Annihilation

An example (case IVa for 111 orientation) is shown in Fig. 7. Here, the threading and misfit dislocations have the same Burgers vector. In the figure, the misfit has moved on its glide plane so as to align anti-parallel to the threading dislocation.

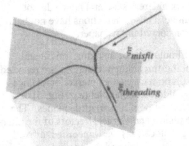

Fig 7. Threading and misfit dislocations having the same Burgers vector align anti-parallel

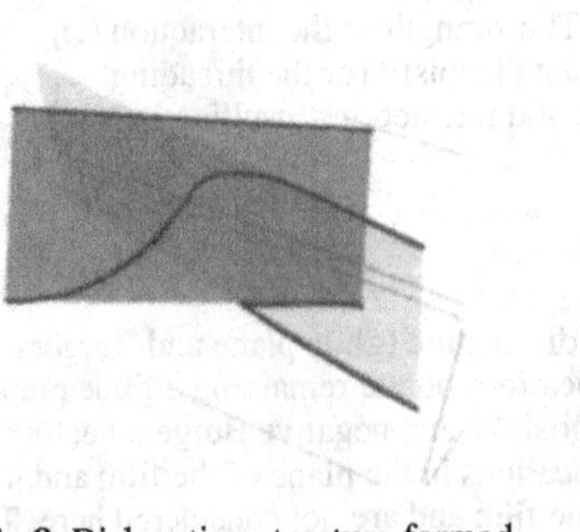

These dislocations subsequently annihilate along the line of intersection of the glide planes and form the structure shown in Fig 8 on unloading. When this dislocation structure was loaded to the same strain again it did not change. A "persistent" dislocation structure formed in all cases where the interaction resulted in annihilation. In all other cases, dislocations separated and returned to

Fig 8. Dislocation structure formed by threading-misfit annihilation

their original configurations on unloading. Using the value for critical stress for a single dislocation σ_{ch} from our simulations as reference, the range of interaction strengths was

$$1.004 \leq \frac{\sigma_{int}}{\sigma_{ch}} \leq 1.298$$

Threading-threading interactions

Since the threading misfit interactions were the same for 111 and 100 orientations, we only looked at 111 oriented films for threading-threading interactions:

Threading dislocations on intersecting planes

Following the reasoning presented for threading-misfit cases, for a given threading dislocation there are four possible threading dislocations on intersecting glide planes that it could interact with. For threading dislocations with the same Burgers vectors, the threading segments annihilate, leaving misfits at both interfaces. Results for the remaining 3 cases are summarized in Table III

Table III Threading dislocations on intersecting planes

Threading	Threading	Interaction	σ_{int}/σ_{ch}
$(1\bar{1}1)[011]$	$(\bar{1}\bar{1}1)[10\bar{1}]$	No reaction	1.113
$(1\bar{1}1)[011]$	$(\bar{1}11)[110]$	Junction formation	3.189
$(1\bar{1}1)[011]$	$(\bar{1}11)[101]$	Junction formation	23.961

Threading dislocations on parallel planes

Two threading dislocations on parallel glide planes could have the same or different Burgers vectors. Fig. 9 shows the variation of strength of threading-threading interaction as a function of spacing between the glide planes.

DISCUSSION

These simple simulations of pairwise dislocation interactions have yielded some interesting results, which may help to explain some phenomena in thin film plastic deformation. Most existing models attribute the high yield stresses observed in thin metal films to the blocking effect of an existing misfit on the motion of a threading dislocation [4, 7, 13]. In our simulations, the

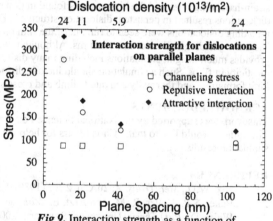

Fig 9. Interaction strength as a function of glide plane spacing for threading dislocations on parallel planes

threading-misfit interaction was found to be at most only 30% stronger than σ_{ch} for 800 nm thick films. For films of the same thickness, threading-threading interactions were found to be stronger than all threading-misfit interactions for plane spacings less than 45 nm. At a plane spacing of 10 nm, interactions between threading dislocations on parallel planes had strength $3.19 \times \sigma_{ch}$. Assuming a one-dimensional periodic array of dislocations and columnar grain structure, a simple calculation for the dislocation density, ρ, is

$$\rho = \frac{2}{s \cdot h} \tag{2}$$

where s is the spacing between dislocations and h is the film thickness. From this relation, dislocations will have 10 nm spacing in an 800 nm thick film when $\rho = 10^{14}/m^2$, which is a reasonable dislocation density for a thin film [8, 11].

It is clear that the stress levels shown in Fig. 9 cannot be interpreted as the strength of the film, since only a fraction of interacting threading dislocations will have spacings smaller than 45 nm and since the presence of additional threading dislocations would reduce the interaction strength of these dipoles. Nonetheless, it is also evident that threading-threading interactions can be much stronger than threading-misfit interactions and may contribute to the high stress levels and high strain hardening rates in thin films. Similar results on the strength of threading-threading interactions were reported by Schwarz [14] for monotonic loading of epitaxial semiconductor films.

The explanation for "negative yielding" proposed by Baker et al. [5] supposes that dislocations "run back" against the applied stress when $\sigma < \sigma_h$. According to our simulation results, only when dislocations annihilate do they not run out of the film on unloading (Fig. 8). All the junctions formed during loading broke on unloading the film. The fact that only 25% of all threading-misfit interactions form permanent dislocation structures supports this explanation of the "negative yield" behavior.

CONCLUSIONS

Results from simulations of pairwise interactions show that threading-threading interactions can be much stronger than threading-misfit interactions. Such interactions may contribute to the high strain-hardening rates and the high stresses found in thin metal films. Annihilation of dislocations resulted in persistent dislocation structures. For all other dislocation reactions, threading dislocations were seen to run out of the film on unloading, which may be the source of "negative yielding" behavior in such films. Although understanding pairwise interactions provides much insight, simulations including many dislocations are required in order to model plasticity in films. Such a simulation should include dislocation interactions resulting in jog and kink formation, and thermally activated climb and cross-slip of dislocations.

ACKNOWLEDGEMENTS

This work was supported by the National Science Foundation under contract DMR-9875119. The authors would like to thank Chris Myers for help with the visualization program for the simulation results.

REFERENCES

1. Nix, W.D., *Met. Trans. A*, **20A**, p. 2217 (1989).
2. Baker, S.P., Keller, R.-M., and Arzt, E., *Mat. Res. Soc. Symp. Proc.*, **505**, p. 605 (1998).
3. Weiss, D., *PhD Thesis*, Universitat Stuttgart, (2000).
4. Freund, L.B., *Advances in Applied Mechanics*, **30**, p. 1 (1994).
5. Baker, S.P., *et al.*, *Mat. Res. Soc. Symp. Proc.*, **516**, p. 409 (1998).
6. Freund, L.B., *J. Appl. Phys.*, **68**(5), p. 2073 (1990).
7. Nix, W.D., *Scripta Mat.*, **39**(4-5), p. 545 (1998).
8. Keller, R.-M., Baker, S.P., and Arzt, E. , *J. Mater. Res.*, **13**(5), p. 1307 (1998).
9. Shu, J., *et al.*, *Mat. Res. Soc. Symp. Proc.*, **563**, p. 207 (1999)
10. Kuschke, W.-M., *et al.*, *J.Mater. Res.* , **13**(10), p. 2962 (1998).
11. Baker, S.P., Kretschmann, A., and Arzt, E., to be published in *Acta Mater.*, 2001.
12. Schwarz, K.W., *J. Appl. Phys.*, **85**(1), p. 108, (1999).
13. Gillard, V.T., Nix, W.D., and Freund, L.B., *J. Appl. Phys.*, **76**(11), p. 7280 (1994).
14. Schwarz, K.W., *J. Appl. Phys.*, **85**(2), p. 120 (1999).

Mat. Res. Soc. Symp. Proc. Vol. 673 © 2001 Materials Research Society

Discrete dislocation simulation of thin film plasticity

B. von Blanckenhagen, P. Gumbsch, and E. Arzt
Max–Planck–Institute for Metals Research, Seestr. 92, D-70174 Stuttgart, Germany

ABSTRACT

A discrete dislocation dynamics simulation is used to investigate dislocation motion in the confined geometry of a polycrystalline thin film. The repeated activation of a Frank–Read source is simulated. The stress to activate the sources and to initiate plastic flow is significantly higher than predicted by models where the dislocations extend over the entire film thickness. An effective source size, which scales with the grain dimensions, yields flow stresses in reasonable agreement with experiments. The influence of dislocations deposited at interfaces is investigated by comparing calculations for a film sandwiched between a substrate and a capping layer with those for a free standing film.

INTRODUCTION

The plastic properties of polycrystalline thin metal films with thickness of $1\mu m$ and below are very different from the properties of the corresponding bulk materials. The flow stresses of thin films can exceed the flow stresses of the bulk materials by an order of magnitude and increase with decreasing film thickness (e. g. [1]). However, the underlying mechanisms are not yet well understood.

Two models are frequently used in the literature to explain the observed high flow stresses of thin films and the dependence on film thickness or grain size. The first balances the energy of dislocations deposited at the interfaces with the work done by the moving dislocation (Nix–Freund model [2, 3]), which results in a flow stress that scales with inverse film thickness. If one requires the deposition of dislocations at the grain boundaries, the flow stress scales with the inverse grain size (Thompson model [4]). Both models qualitatively explain the increase in the flow stress with decreasing film thickness (or grain size), but the predicted flow stresses are often much smaller than the experimentally measured ones [1, 5].

Interface dislocations deposited by dislocation glide have been observed in transmission electron microscopy [6, 7], but are reported not to be very stable and to vanish after a short time of irradiation with the electron beam [8, 9]. Other authors even observe that the interfaces seem to operate as a sink for the dislocations and not as an obstacle [10]. However, in situ transmission electron microscopy studies of thin film plasticity show that dislocation sources operate in the grain interior and that single sources repeatedly emit dislocation loops [10].

The dislocation evolution in a columnar grain of a thin film is simulated here with a discrete dislocation dynamics method. Grain boundaries and interfaces are introduced as impenetrable obstacles for the dislocations. The model geometry corresponds to a thin film deposited on a substrate covered by a capping layer and is contrasted with simulations for a free standing film where the dislocations can leave the film at the surfaces. A pinned dislocation segment is positioned in the grain interior to mimic a Frank–Read source. The

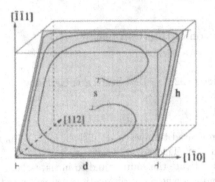

Figure 1: Model geometry for dislocations gliding on a (111) plane in a $[\bar{1}\bar{1}1]$ textured grain.

operation of the source in the confined geometry of the thin film is investigated.

METHOD

The discrete dislocation dynamics simulation used here is optimised for calculation of few dislocations (< 100) with an accurate calculation of the dislocation self–stresses. It is based on linear isotropic elasticity. Curved dislocations are represented by nodes connected by straight segments of arbitrary character. The dynamic is established by moving the nodes according to a viscous drag law. The force on every node is calculated from the applied stress, the stresses to fulfill the boundary conditions, the self interaction and the interaction with other dislocations [11, 12]. Further details of the simulation and test–calculations are described in Ref. [13].

Dislocation movement in a single columnar grain is investigated. Only one active dislocation source is considered and positioned in the centre of the grain. The source consists of a dislocation pinned at two points which are separated by a distance s (see figure 1). The arms of this source on a secondary glide plane are neglected. The Burgers vector of the dislocations $\mathbf{b} = \frac{1}{2}a_0[1\bar{1}0]$ (a_0 being the lattice parameter) is parallel to the film–substrate interface. These initial conditions were chosen to make the activation of the source as easy as possible in order to get a lower bound for the influence of the source on the flow stress: A Frank–Read source with an initial edge segment has the lowest activation stress. The back stress of the dislocations at the boundaries on the source will be lowest if the source is in the centre of the grain. The calculations were performed for rectangular grains where the film thickness projected onto the glide plane, h, equals the grain size, d.

The glide plane is confined by grain boundaries to the left and to the right, which are introduced as impenetrable obstacles (figure 1). Transmission of dislocations through the grain boundaries or initiation of plastic flow in a neighbouring grain has been considered in a previous paper [13] and is not dealt with here. A free standing film and a film sandwiched between a substrate and passivation layer are simulated. In the latter case the interfaces to substrate and passivation layer are treated also as impenetrable obstacles. Differences between the elastic moduli of film, capping–layer and substrate, which would lead to image forces, are neglected. This approximation is justified because most of the dislocations are located in the grain interior far from the boundaries and the self–stress due

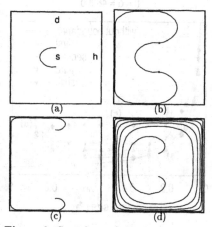

Figure 2: Snapshots of the simulation for special points in figure 3.

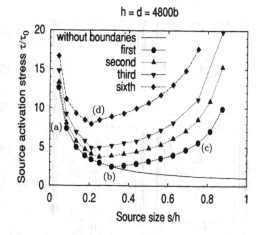

Figure 3: Source activation stress in units of the stress needed to activate a source of size $s = d$ versus size of the source normalised by the film thickness.

to the curvature of the dislocations is usually much larger than the image forces.

In a free standing film the dislocations can leave the film at the surfaces. A dislocation which touches the surface is simply terminated at the free surface. The terminating nodes are displaced in such a way that the last segment of the dislocation enters the free surface at a right angle. This is a good approximation for an edge dislocation. For mixed dislocations there is a tendency to align in the direction of the Burgers vector to minimise line energy [12]. However, as long as the critical process for moving the dislocations is the activation of sources in the grain interior, the treatment of the dislocations close to the surfaces plays only a minor role in determining the flow stress. The importance of image forces in the grain interior was estimated by comparison of calculations for a free standing film with and without the correct boundary conditions for the free surfaces. To establish the boundary conditions of the free surfaces, point forces were applied on the surface [14] to compensate for the dislocation induced tractions. Interaction between the two free surfaces has also been taken into account following Ref. [15]. The calculations with and without the image forces agreed within a few percent. The image forces are therefore neglected in the following simulations of free standing films.

RESULTS

The source activation stress is defined as the maximum stress needed to create a complete loop and to restore the original source configuration. It is plotted for the generation of subsequent dislocation loops in figure 3 in units of the stress τ_0 to activate a Frank–Read source of size $s = d$. The lowest curve in figure 3 corresponds to the source activation stress for the first loop. For small source sizes (point (a) in figure 2 and 3) the activation stress is high due to the large curvature of the dislocation in its critical configuration. The source activation stress decreases for larger sources and follows the solid

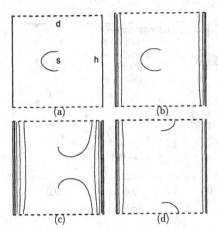
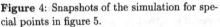

Figure 4: Snapshots of the simulation for special points in figure 5.

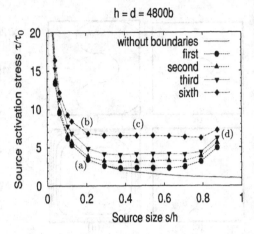

Figure 5: Source activation stress in units of stress τ_0 versus size of the source normalised by the film thickness for a free standing film.

line in figure 3 if the boundary constraints are neglected and only the activation of the Frank–Read source is considered. With boundary conditions which confine the dislocations within the grain the stress to produce a dislocation loop increases again when the source becomes larger than one third of the film thickness. In this regime, the distance between a pinning point and the interface becomes smaller than s and the critical configuration changes from activating the source to the passage of the dislocation between the source and the interface (figure 2(b) and (c)). The Frank–Read source then changes into a double spiral source. With increasing number of dislocations increasing stress is required to produce additional loops (figure 2(d)) and the source size which has the lowest activation stress shifts to smaller values. For the first dislocation the source with the lowest activation stress has a size of $s = d/3$. If there are already five loops at the boundaries it is approximately $s = d/5$.

Figure 5 shows the source activation stress in a free standing film. For small sources (figure 4(a) and (b)) the stresses are similar to the activation stresses for the sandwiched film but the increase in activation stress with increasing number of loops is somewhat smaller (compare figure 3 and 5). For sources larger than $d/3$ the activation stress reaches a plateau value. Here, the critical configuration is the passage between the pileup at the boundary and the pinning point (figure 4(c)). The stress value of the plateau region for the first dislocation is somewhat larger than for the unconstrained Frank–Read source because the dislocation in configuration 4(c) corresponds to a source with an initially straight segment of screw character and requires a higher stress to bow out than configuration 4(a) with edge character.

If the source size becomes larger than $\frac{3}{4}d$, the stress increases again. Now a Frank–Read source which is formed by the dislocation terminating at the boundary and its image dislocation has to be activated (figure 4(d)). In this size regime the simulation can only give a qualitative determination of the activation stresses, since image forces are not correctly included. The activation stresses will also be underestimated for the critical configuration 4(c): The dislocations in the pile-ups will be closer together at the surfaces since the back stress of the image dislocations is neglected. This results in a slight decrease of the activation stress as s becomes larger than $d/2$. However, the inclusion of image forces only increases the source activation stress, and the results can therefore be regarded as a lower bound.

DISCUSSION

Our results for a film on a substrate and with passivation layer (sandwiched configuration) show that the size of a dislocation source which is most efficient in producing several dislocation loops, i. e. which has the lowest activation stress, is roughly $s \approx d/4 = h/4$ (figure 3). For aspect ratios of film thickness to grain size different from one, the smaller of both will control the flow stress, since the dominant obstruction to the dislocation motion is given by the smallest distance between pinning points and interfaces or boundaries.

Smaller sources require a higher stress to operate. Larger sources initially bow out at a lower stress, but are then obstructed by the boundaries. If several dislocation sources of different sizes are present in a grain, sources with a size of roughly one quarter of the grain dimensions, $s^{\text{eff}} \approx d/4$, will produce dislocation loops most easily and will first be activated. Since the stress to operate a source of size s is, neglecting logarithmic terms, given by $\tau_{\text{source}} = \mu b/s$, where μ is the shear modulus, the stress to initiate plastic flow is inversely proportional to the grain dimensions. Furthermore, with $s^{\text{eff}} \approx d/4$ the flow stress is roughly four times higher than predicted by the Nix or Thompson model where the dislocations extend over the entire film thickness or grain size. The agreement with experiment is also much better than for these models. The flow stresses determined from $s^{\text{eff}} \approx d/4$ are comparable to the measured ones (e.g. [5]).

Only small changes are expected if substrate and capping layer are stiffer than the film material: For the case of one dislocation loop in the film it was shown that the stand-off distance between the dislocation and the interface to a stiffer layer is rather small [16]. Furthermore, the stress needed to move a dislocation in a channel with an elastically rigid substrate and capping layer is only 20% higher than for a channel where substrate, capping layer and film have the same elastic constants [17]. Since the image forces are largest for the dislocations closest to the interface, the effect should be even smaller for subsequent dislocation loops further inside the grain.

The results for the free standing film are qualitatively similar to the sandwiched configuration. Small and very large sources require a high stress to operate. A most efficient source size can be deduced. This size is determined by the transition of the critical configuration from figure 4(a) to 4(c). For the first activation it is approximately $s^{\text{eff}} \approx d/3$ and becomes smaller for the successive loops. The most efficient size will be slightly smaller for a different Burgers vector orientation (the movement of the dislocation past configuration 4(c) becomes easier) and larger for a different source orientation (the pinning points come closer to the grain boundary). Nevertheless, the flow stress is at least half the flow stress for the sandwiched film and twice as high as predicted by the Thompson model.

CONCLUSION

In these discrete dislocation simulations it was assumed that dislocation sources in thin films are rare and operate several times to achieve plastic deformation of a thin film. The activation of dislocation sources then is the most difficult step in the deformation process. A flow stress is found which scales with the inverse grain size or film thickness and which is roughly four times higher than predicted by previous models.

The flow stress of a free standing film is controlled by the grain size. The dislocations can leave the film at the interfaces, nevertheless, the magnitude of the flow stress is at least half that of a sandwiched film with the same grain size.

REFERENCES

[1] R.-M. Keller, S. P. Baker, and E. Arzt, J. Mater. Res. **13**, 1307 (1998).

[2] W. D. Nix, Metall. Trans. A **20A**, 2217 (1989).

[3] L. B. Freund, J. Appl. Mech. **54**, 553 (1987).

[4] C. V. Thompson, J. Mater. Res. **8**, 237 (1993).

[5] O. Kraft, M. Hommel, and E. Arzt, Mater. Sci. Eng. **A288**, 209 (2000).

[6] R. Venkatraman, J. C. Bravman, W. D. Nix, P. W. Davies, P. A. Flinn, and D. B. Fraser, J. Electron. Mater. **19**, 1231 (1990).

[7] D. Jawarani, H. Kawasaki, I.-S. Yeo, L. Rabenberg, J. P. Start, and P. S. Ho, J. Appl. Phys. **82**, 171 (1997).

[8] R. Venkatraman, S. Chen, and J. C. Bravman, J. Vac. Sci. Technol. A **9**, 2538 (1991).

[9] P. Müllner and E. Arzt, Mater. Res. Symp. Proc. **505**, 149 (1998).

[10] G. Dehm and E. Arzt, Appl. Phys. Lett. **77**, 1126 (2000).

[11] B. Devincre and M. Condat, Acta Metall. Mater. **40**, 2629 (1992).

[12] K. W. Schwarz, J. Appl. Phys. **85**, 108 (1999).

[13] B. von Blanckenhagen, P. Gumbsch, and E. Arzt, Modelling Simul. Mater. Sci. Eng. **9**, 157 (2001).

[14] M. C. Fivel, T. J. Gosling, and G. R. Canova, Modelling Simul. Mater. Sci. Eng. **4**, 581 (1996).

[15] A. Hartmaier, M. C. Fivel, G. R. Canova, and P. Gumbsch, Modelling Simul. Mater. Sci. Eng. **7**, 781 (1999).

[16] J. D. Embury and J. P. Hirth, Acta Metall. Mater. **42**, 2051 (1994).

[17] W. D. Nix, Scripta Metall. **39**, 545 (1998).

Mat. Res. Soc. Symp. Proc. Vol. 673 © 2001 Materials Research Society

Influence of Film/Substrate Interface Structure on Plasticity in Metal Thin Films

G. Dehm, B.J. Inkson*, T.J. Balk, T. Wagner, and E. Arzt
Max-Planck-Institut für Metallforschung, Seestr. 92, 70174 Stuttgart, Germany
*Department of Materials, University of Oxford, Parks Road, Oxford, OX1 3PH, U.K.

ABSTRACT

In-situ transmission electron microscopy studies of metal thin films on substrates indicate that dislocation motion is influenced by the structure of the film/substrate interface. For Cu films grown on silicon substrates coated with an *amorphous* SiN_x diffusion barrier, the transmission electron microscopy studies reveal that dislocations are pulled towards the interface, where their contrast finally disappears. However, in epitaxial Al films deposited on single-crystalline α-Al_2O_3 substrates, threading dislocations advance through the layer and deposit dislocation segments adjacent to the interface. In this latter case, the interface is between two *crystalline* lattices. Stresses in epitaxial Al films and polycrystalline Cu films were determined by substrate-curvature measurements. It was found that, unlike the polycrystalline Cu films, the flow stresses in the epitaxial Al films are in agreement with a dislocation-based model.

INTRODUCTION

The stresses that develop in metal thin films deposited on rigid substrates can significantly exceed the flow stresses in the corresponding bulk metal. This phenomenon has been attributed to geometrical constraints on the films, which alter the energetics of dislocation motion. Dislocation-based models for the flow stress assume that the film/substrate interface acts as an obstacle for dislocations moving in the metal film. A dislocation gliding on a plane inclined to the film/substrate interface creates a dislocation line at this interface [1,2].

A number of transmission electron microscopy (TEM) studies have been performed in order to investigate dislocation plasticity in metal thin films [3-13]. While some TEM studies report the formation of interfacial dislocations during straining of metal thin films [3-5], there are also TEM observations describing the fading of interfacial dislocation contrast due to disappearance of their strain field [5-7]. However, unstable interfacial dislocations contradict the concept of dislocation-based models for the flow stress. Furthermore, work hardening in thin films, which is explained by the interaction of gliding dislocations with interfacial dislocation segments, would have to be reconsidered. Recently, the high flow stresses in thin Ag films were ascribed to thermally activated dislocation motion [8]. TEM observations revealed smooth dislocation motion in Ag grains at elevated temperatures (low film stresses), but jerky dislocation motion at low temperatures (high film stresses), with pinning distances significantly smaller than the grain dimensions [8].

In the present study, we have investigated dislocation-interface interactions and dislocation motion in two metal thin film systems with different film/substrate interface structure. Polycrystalline Cu films were deposited on amorphous SiN_x (a-SiN_x) coated (001) Si substrates, and epitaxial Al films were grown on (0001) α-Al_2O_3 substrates. The flow stresses of the polycrystalline Cu films and epitaxial Al films will be discussed in conjunction with the TEM observations of dislocation behavior in those films.

EXPERIMENTAL DETAILS

Film deposition

{111} textured Cu films were deposited by magnetron sputtering in an ultra-high vacuum (UHV) environment (base pressure: 2×10^{-7} Pa). As substrate material, 500µm thick (001) oriented Si wafers coated with an amorphous SiO_x (a-SiO_x) layer and an a-SiN_x diffusion barrier were used. Both amorphous interlayers were nominally 50nm thick. The substrate surface was cleaned prior to Cu deposition by a 200eV Ar-ion bombardment for 3 minutes. 99.999% pure Cu was deposited at room temperature onto the rotating substrate (100rev/min) at a growth rate of 75nm/min (power: 200W). The Cu films, which were annealed at 600°C for 10min directly after deposition without breaking vacuum, possessed a {111} texture.

Epitaxial Al films with thicknesses ranging from 200nm to 2000nm were magnetron sputtered onto (0001) oriented α-Al_2O_3 single crystals in a UHV system with a base pressure of better than 2×10^{-8} Pa. The 50mm diameter α-Al_2O_3 substrates (thickness 330µm) were ultrasonically cleaned in acetone and ethanol. Subsequently, the substrates were Ar-ion sputter cleaned (Ar-pressure 10^{-2} Pa, kinetic energy 200eV) in the UHV system and annealed at 1000°C for 1 hour in order to remove structural defects. Magnetron sputtering was carried out at ambient temperature in 99.9999% pure Ar. During magnetron sputtering, the temperature of the Al films rose to ~70°C. No contaminants were detected by Auger-electron spectroscopy on the surface of the freshly deposited Al films. In addition, a 350nm thick epitaxial Al film was grown by molecular beam epitaxy (MBE) onto a (0001) oriented α-Al_2O_3 single crystal with dimensions 10mm $\times$ 10mm $\times$ 0.5mm. All Al films grew epitaxially as two twin variants with the orientation relationships (111) Al $\parallel$ (0001) α-Al_2O_3 and $\pm[\bar{1}10]$ Al $\parallel$ [$10\bar{1}0$] α-Al_2O_3 [14].

Thermal stress measurements

Stresses in the films arising from differences in thermal expansion between the film and the substrates were measured using a laser scanning technique [15]. The epitaxial Al films were thermally cycled between room temperature (RT) and a maximum temperature of 400°C, while the Cu films were cycled between RT and 500°C. The cooling and heating rates were maintained at 6°C/min, and the furnace was purged with nitrogen before and during thermal cycling. The biaxial film stress σ_f was calculated from the substrate-curvature data at a given temperature using the Stoney equation [16]:

$$\sigma_f = \frac{M_s h_s^2}{6 h_f} \frac{1}{R} \tag{1}$$

where R is the radius of curvature, $M_s = E_s/(1-v_s)$ is the biaxial elastic modulus of the substrate, E_s and v_s denote the Young's modulus and Poisson's ratio of the substrate, and h_f and h_s are the thicknesses of the film and the substrate, respectively.

In-situ TEM studies and TEM specimen preparation

In-situ thermal cycling experiments of Al films were performed using a 400keV TEM (Jeol 4000FX), while Cu films were studied using a 1 MeV high-voltage TEM (AEI/EM7 [17]). Both TEMs are equipped with double-tilt heating stages and TV-rate cameras. The TEM specimens were thermally cycled under vacuum at a column pressure of less than 10^{-4}Pa. The temperature was ramped manually in steps of 10 to 50°C/min to maximum temperatures of 400°C for Al films and 600°C for Cu films.

Detailed in-situ TEM experiments were performed on a nominally 350nm thick MBE grown Al film and a nominally 500nm thick Cu film using plan-view and cross-sectional TEM specimens. TEM measurements of the Cu film in cross-section revealed that the thickness was actually 450nm. Al/Al_2O_3 plan-view TEM foils were prepared by boring disks 3mm in diameter along the [0001] α-Al_2O_3 direction. The disks were mechanically thinned, dimpled and Ar-ion milled at 3-6kV from the α-Al_2O_3 rear side until electron transparency was achieved. Cross-sectional Al/Al_2O_3 specimens were sandwiched to itself, glued into a ceramic tube, cut in slices, dimpled, and Ar-ion milled following the routine described in [18]. In contrast, cross-sectional TEM specimens of the Cu/a-SiN_x/a-SiO_x/Si system were made using a focused ion beam (FIB) microscope (FEI 200). Strips of 3mm length and ~100µm thickness were glued to a Cu half-ring, and electron transparent regions ~30µm long, 2-3µm deep, and ~400nm thick along the electron beam direction were milled with a focused Ga-ion beam [19]. Plan-view Cu/a-SiN_x/a-SiO_x/Si specimens were mechanically thinned and dimpled from the Si backside. Finally, the dimpled surface of the Si substrate was etched with HF/HNO_3. The a-SiO_x/a-SiN_x bilayer acted as an etch stop, and thus the Cu film remained completely intact even in the electron transparent region of the specimen.

RESULTS

Polycrystalline Cu films

Substrate-curvature measurements of a nominally 400nm thick Cu film revealed biaxial compressive stresses of ~130MPa at 500°C, while during cooling from 500°C to RT, biaxial tensile stresses of more than 570MPa were generated. Stress-temperature curves for the first two thermal cycles of the film are shown in Figure 1.

Dynamic dislocation observations were performed on cross-sectional TEM specimens with two-beam diffraction conditions, for which the g_{111} diffraction vector was perpendicular to the film/substrate interface and lay between $<11\bar{2}>$ and $<01\bar{1}>$ Cu zone-axes. The in-situ observations revealed that dislocations advancing through the film did not deposit interfacial dislocation segments and were not hindered in their motion by the Cu/a-SiN$_x$ interface independent of the temperature. On the contrary, the Cu/a-SiN$_x$ interface was frequently observed to attract dislocations and to act as a dislocation sink. This behavior is demonstrated in Figure 2, which shows a sequence of fixed images of a video film recorded during a heating cycle. Dislocations with a projected line direction approximately parallel to the interface were pulled towards the Cu/a-SiN$_x$ interface and accelerated in motion until their contrast finally disappeared (Figs. 2a-c). Dislocation loops, which formed in the center of a grain [11], expanded continuously in diameter. The segments of the dislocation loop approaching the interface accelerated for a short period of time and subsequently vanished upon contact with the interface (Figs. 2c,d). This behavior was observed for several dislocation loops expanding in that grain over a temperature range of 380°C to 420°C. During cooling, the interaction of dislocations with the Cu/a-SiN$_x$ interface was the same as at elevated temperatures. However, dislocation motion changed gradually, from continuous to jerky, with decreasing temperature. Below 200°C, dislocation motion in cross-sectional TEM specimens nearly ceased, and the thermal stress increased upon cooling (see Fig. 1) at a higher rate than during the initial stage of cooling. No buckling of the FIB-prepared TEM cross-sectional specimen was noticed during thermal cycling.

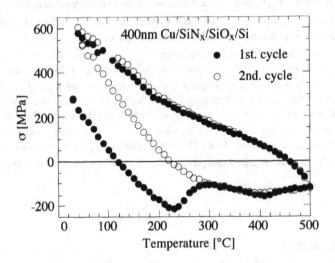

Figure 1. Stress evolution in a nominally 400nm thick Cu film on a-SiN$_x$ coated Si during the first and second thermal cycles.

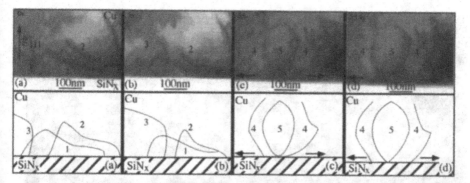

Figure 2. Sequence of cross-sectional TEM images recorded at ~410°C during heating of a 450nm thick Cu film deposited on a Si substrate with an a-SiN$_x$ diffusion barrier. (a,b) Dislocations 1 and 2 are pulled towards the Cu/a-SiN$_x$ interface, while dislocation 3 moves through the Cu grain without being hindered by the interface. (c,d) Dislocation loops 4 and 5 expand in the film. Dislocation loop segments are attracted to the Cu/a-SiN$_x$ interface.

Plan-view TEM studies of thermal stress induced dislocation motion are in agreement with the observations made on the cross-sectional specimens. Deposition of dislocation segments at the Cu/a-SiN$_x$ interface was not observed. However, the configurations, motion and density of dislocations changed during thermal cycling. At elevated temperatures, most dislocations extended over a large fraction of the grain diameter (Fig. 3a), independent of the stress state of the film (compressive or tensile). In contrast, cooling to temperatures below ~220°C resulted in significantly shorter dislocation segments that formed tangles (Fig. 3b). Mobile dislocations were present at all temperatures. However, dislocation motion was thermally activated, causing a transition from *continuous* to *jerky* motion upon cooling, as was also observed in the cross-sectional TEM specimens. At elevated temperatures, dislocations move over distances up to the grain diameter, while with decreasing temperature the mean free path between pinning sites decreases. Figure 4 shows the measured distances that dislocations moved during cooling from 185°C to 155°C. The mean free path of dislocation motion in this temperature regime is ~70nm (median value: ~45nm). In contrast, during cooling from 470°C to 440°C, dislocations advanced ~260nm on average. The dislocation densities remained rather constant, with values of $(3.4\pm0.8)\cdot10^9$cm^{-2} for elevated temperatures (600°C to 440°C), and $(5.9\pm0.6)\cdot10^9$cm^{-2} for temperatures below 230°C. Even after 3 cycles, no noticeable increase in dislocation density was detected. Stress evolution similar to that presented in Figure 1 is expected to exist in the Cu film of the plan-view TEM specimen, since it did not buckle due to the underlying a-SiN$_x$, a-SiO$_x$ and some residual Si, which was detected by selected area diffraction.

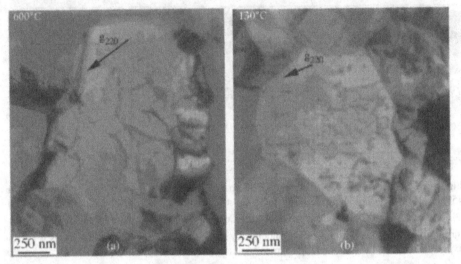

Figure 3. Plan-view TEM images of dislocations in Cu grains at (a) 600°C and (b) 130°C. No interfacial dislocations deposited by advancing threading dislocations are discernable in the images recorded close to <111> Cu zone axes. At elevated temperatures, (a) dislocations appear to be longer and more mobile than at lower temperatures (b), where dislocation motion became jerky and dislocation tangles formed.

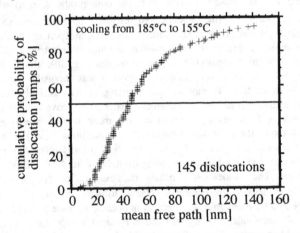

Figure 4. Distribution of distances that dislocations moved between pinning sites. The values were measured using frame by frame inspection of the video film recorded during the 2ⁿᵈ and 3ʳᵈ temperature cycles during cooling (T=185°C...155°C). Only very few dislocations glide over distances comparable to the average grain size of 620nm at a film thickness of 450nm, while most dislocations move distances which are only a fraction of the grain dimensions.

Epitaxial Al films

Figure 5 shows a cross-sectional TEM image of a 350nm thick epitaxial Al film on (0001) α-Al_2O_3 taken with a g_{111} two-beam diffraction condition perpendicular to the film/substrate interface and close to a $<11\bar{2}>$ Al zone-axis. The micrograph shows a threading dislocation which deposited a 60° dislocation segment nearly parallel to the (111)Al/(0001) α-Al_2O_3 interface. The dislocation advanced through the Al film upon localized heating with a focused electron beam. The heating is caused by radiation from the objective aperture, which absorbs diffracted electrons, and by electron-specimen interactions. Dislocation glide occurred on a $(1\bar{1}1)$ Al plane, which is inclined 70.5° to the Al/α-Al_2O_3 interface. Note that the Al film/native oxide interface is devoid of a discernable interfacial dislocation segment.

Threading dislocations advancing through the film and dragging behind a dislocation segment parallel to the Al/α-Al_2O_3 interface were also observed during in-situ TEM experiments using a heating-stage. Dynamic dislocation studies revealed that the dislocation end terminating near the film/substrate interface was less mobile than the end at the film/native oxide interface, since its motion was hindered by an Al/α-Al_2O_3 interfacial dislocation network. However, the Al/(0001)α-Al_2O_3 interface also acted as a dislocation source during thermal cycling. Emission of dislocation half-loops from the Al/α-Al_2O_3 interface has frequently been observed, while this was not the case for the Cu/a-SiN_x interface.

Plan-view TEM studies of the 350nm thick Al film revealed dislocations which were more than 1.5µm long, indicating that they lie parallel to the Al/(0001)α-Al_2O_3 interface. The contrast of these dislocations did not disappear under electron irradiation. The dislocation density at the end of a thermal cycle was measured to be $4\times10^9 cm^{-2}$ under $g_{11\bar{1}}$ two-beam conditions near a $<112>$ Al zone axis. However, it must be noted that plan-view samples were less instructive than cross-sectional specimens for the observation of dislocation activity, due to the existence of small holes in the samples that were induced by ion-milling. Moreover, the Al film was not directly supported by a substrate in some small localized regions, and buckling of the plan-view TEM sample occurred during thermal cycling.

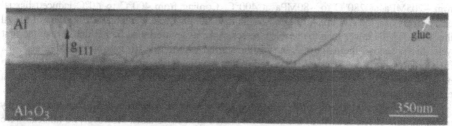

Figure 5. Cross-sectional TEM image of a 350nm thick epitaxial Al film deposited on a single-crystal (0001) α-Al_2O_3 substrate. A threading dislocation has dragged behind a dislocation segment that is at stand-off from the Al/α-Al_2O interface. Note that this dislocation segment interacts with a dislocation network already present at the interface. Furthermore, dislocation loops can be seen bowing out of the interface.

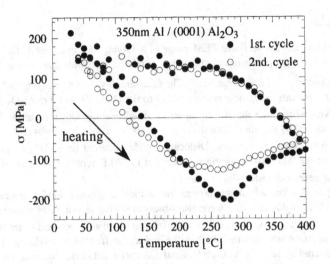

Figure 6. Stress evolution during the first and second thermal cycles of a nominally 350nm thick epitaxial Al film that had been sputter deposited onto (0001) α-Al$_2$O$_3$.

Figure 6 shows the first and second stress-temperature curves of a sputter deposited 350nm thick epitaxial Al film on (0001) α-Al$_2$O$_3$. At the start of the first stress-temperature cycle, the as-sputtered Al film was under a biaxial tensile stress of ~210MPa. During heating the tensile stresses in the Al film relaxed along a thermo-elastic line and became biaxial compressive stresses, as a result of the constraint from the α-Al$_2$O$_3$ substrates and the difference in thermal expansion coefficients between Al ($\alpha_{Al} \approx 23 \times 10^{-6}K^{-1}$) and α-Al$_2O_3$ ($\alpha_{Al2O3} \approx 7 \times 10^{-6}K^{-1}$). The slope of the thermo-elastic curve was measured to be 1.8 MPa/K, in agreement with a theoretically calculated slope of 1.85MPa/K. When heated further, the as-sputtered Al film started to yield, and a stress drop occurred at temperatures exceeding 280°C. The compressive stresses decreased from ~205MPa at 280°C to ~80MPa at 400°C. Cooling from 400°C to 40°C induced biaxial tensile stresses in the Al film. At the end of the first thermal cycle, the tensile flow stress of the Al film was ~70MPa lower than the tensile stresses of the as-sputtered Al film at the start of the first cycle. In the second thermal cycle, the Al film started with tensile stresses of ~120MPa at 55°C. The compressive flow stresses decreased rather continuously from ~130MPa at ~270°C to ~60MPa at 400°C without a pronounced yield drop. At the end of the second thermal cycle, a tensile flow stress of ~150MPa was measured (Fig. 6).

Scanning electron microscopy (SEM) and FIB studies of as-sputtered Al films and thermally cycled Al films reveal growth of the two twin related growth variants. However, no hillocks were observed by SEM and FIB, even after several heat treatments. Similar findings were reported for MBE grown Al films on α-Al$_2$O$_3$ [14].

The flow stresses of the epitaxial Al films with thicknesses ranging from 200nm to 2000nm are summarized in Table 1. The tensile flow stresses at 40°C were measured at the end of the first thermal cycle.

Table 1. Tensile flow stresses of the epitaxial Al films measured at the end of the first thermal cycle after cooling from 400°C to 40°C.

thickness / flow stress	200 nm	350 nm	600 nm	1000 nm	2000 nm
σ (40°C)	220 MPa	150 MPa	120 MPa	109 MPa	60 MPa

DISCUSSION

The TEM studies of Cu/a-SiN$_x$/a-SiO$_x$/Si and Al/α-Al$_2$O$_3$ specimens indicate that the nature of the film/substrate interface influences dislocation mechanisms in thermally strained thin films on substrates. For the case of an interface between two crystalline materials, such as Al films on α-Al$_2$O$_3$ substrates or semiconductor films on semiconductor substrates [e.g. 20], dislocations have been observed to deposit dislocation segments adjacent to the interface, as assumed in the flow stress model of Nix [1]. However, in the case of an interface between a metal film and an amorphous substrate, the present TEM results imply that interfacial dislocation segments are unstable, with the interface acting as a dislocation sink. The different behavior of the Cu/a-SiN$_x$ and Al/α-Al$_2$O$_3$ interfaces may be a result of their atomic structure and bonding interactions. At a crystalline/amorphous interface, the lack of periodicity in the atoms of the amorphous material leads to an incoherent interface structure (Fig. 7a) with dangling bonds due to a locally varying mismatch of adjacent atoms across the interface. If now a threading dislocation advancing through the film terminates at the crystalline/amorphous interface, it appears to be energetically more favorable to locally relax interfacial atomic positions than to maintain most of the interfacial atomic positions with their dangling bonds and simultaneously to create an additional interfacial dislocation segment. Furthermore, the roughness of 10-20Å of the Cu/a-SiN$_x$ interface,

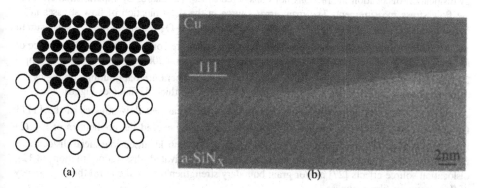

(a) (b)

Figure 7. (a) Schematic drawing of a rigid lattice model of an interface between a crystalline (filled circles) and amorphous material (open circles). (b) High resolution TEM image of the Cu/a-SiN$_x$ interface, revealing a roughness of several (111) Cu lattice planes at the interface.

as observed by high resolution TEM (Fig. 7b), may assist in relaxing the strain field of dislocations that approach the interface and/or of atomic steps created by dislocations running into the interface. In contrast, the Al/α-Al$_2$O$_3$ interface is atomically flat [14], and terraces of e.g. (0006) α-Al$_2$O$_3$ lattice planes (~2.16Å) in height form. Similar to the above mentioned dislocation-interface interactions, it has been reported in the literature that the atomic structure, bonding and stiffness of heteroepitaxial interfaces between two crystalline materials control the stand-off position of misfit dislocations [21-23] and the misfit dislocation core structure [24,25].

The TEM observations of epitaxial Al films on α-Al$_2$O$_3$ mirror the deformation mechanisms upon which the flow stress model of Nix is based [1]. Consequently, the flow stresses for different film thickness should follow the predictions of the Nix model [1]. Figure 8 presents the tensile flow stresses of the epitaxial Al films, measured after cooling from 400°C to 40°C (Table 1), as a function of the inverse film thickness, and the corresponding flow stresses calculated using the Nix model:

$$\sigma_{Nix} = \frac{\sin\phi}{\cos\phi\cos\lambda} \frac{b}{2\pi(1-\nu)} \frac{1}{h} G_{eff} \text{ with } G_{eff} = \frac{G_f G_s}{G_f + G_s} \ln\left(\frac{\beta_s h}{b}\right) \qquad (2)$$

where $\phi = 70.5°$ is the angle between the epitaxial (111) Al film normal and the $\{1\bar{1}1\}$ glide plane normal, $\lambda = 35.3°$ the angle between the film normal and the Burgers vector $\mathbf{b} = 1/2 <110>$, b = 2.86Å the magnitude of the Burgers vector, $\nu = 0.36$ the Poisson's ratio of Al, h the Al film thickness, $G_f = 26$GPa and $G_s = 185$GPa the shear moduli of the Al film and the α-Al$_2$O$_3$ substrate, respectively, and $\beta_s = 1.7$ a numerical constant defining the cutoff radius of the stress field of the interfacial dislocation. The calculated flow stresses are in agreement with the experimentally determined flow stresses (Fig. 8). The slightly larger values of experimental flow stress could be caused by the presence of twin boundaries separating the two Al growth domains, by dislocation-dislocation interactions not considered in Eq. (2) and/or by experimental errors in the flow stress measurement. The main error source in the determination of flow stresses in Al films on α-Al$_2$O$_3$ lies in the curvature value K of the bare α-Al$_2$O$_3$ substrate, which changes up to $\Delta K \approx 0.002$m^{-1} during a thermal cycle. This change in curvature combined with the large value of $M_s = 576$GPa results in a worst case error of 10MPa for $h_f = 2000$nm and 100MPa for $h_f = 200$nm. Moreover, the scatter of the experimental stress-temperature data during cooling cycles demonstrates the uncertainties in the stress values for a 350nm thick Al film (see Fig. 6).

In contrast to the epitaxial Al films on α-Al$_2$O$_3$, the measured flow stresses of the $\{111\}$ textured Cu films on a-SiN$_x$ coated Si substrates significantly exceed the predictions of Eq. (2) [e.g. 15,26]. For this system, the present study reveals that no localised interfacial dislocations form. Therefore, other mechanisms such as thermally activated dislocation motion [8,12], dislocation source effects [27] and/or grain boundary strengthening could control the dependency of flow stress on film thickness.

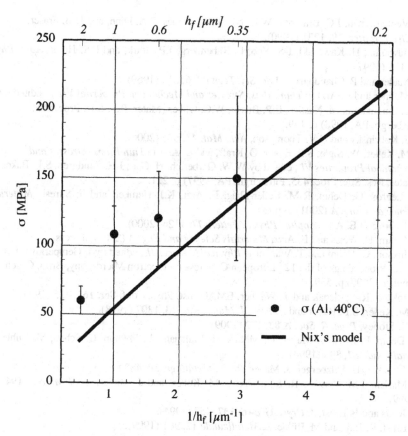

Figure 8. Flow stress for epitaxial Al films measured at 40°C at the end of the first temperature cycle, as a function of reciprocal film thickness. The film thickness is indicated on the upper scale. The heavy line represents the predictions of Eq. (2) (Nix model) using the following values: $\sin\phi/\cos\phi \cos\lambda = 3.46$, $G_f = 26GPa$, $G_s = 185GPa$, $\beta_s = 1.7$, and $\nu = 0.36$.

ACKNOWLEDGEMENT

Financial support by the Deutscher Akademischer Austauschdienst (DAAD), The British Council and The Royal Society is gratefully acknowledged.

REFERENCES

1. W.D. Nix, *Metall. Trans. A* **20**, 2217 (1989).
2. W.D. Nix, *Scripta Mater.* **39**, 545 (1998).
3. T.S. Kuan and M. Murakami, *Metall. Trans. A* **13**, 383 (1982).

4. R. Venkatraman, J.C. Bravman, W.D. Nix, P.W. Davies, P.A. Flinn, and D.B. Fraser, *J. Electr. Mater.* **19**, 1231 (1990).

5. D. Jawarani, H. Kawasaki, I-S. Yeo, L. Rabenberg, J.P. Stark, and P.S. Ho, *J. Appl. Phys.* **82**, 171 (1997).

6. S. Mader and P. Chaudhari, *J. Vac. Sci. Technol.* **6**, 615 (1969).

7. P. Müllner and E. Arzt in *Thin Films: Stresses and Mechanical Properties VII* , edited by R.C. Cammarata, M. Nastasi, E.P. Busso, W.C. Oliver (Mater. Res. Soc. Proc. 505, Pittsburgh, PA, 1998) p. 149.

8. M.J. Kobrinsky and C.V. Thompson, *Acta Mat.* **48**, 625 (2000).

9. R-M. Keller, W. Sigle, S.P. Baker, O. Kraft, and E. Arzt in *Thin Films: Stresses and Mechanical Properties VI* , edited by W.W. Gerberich, H. Gao, J-E. Sundgren, S.P. Baker (Mater. Res. Soc. Proc. 435, Pittsburgh, PA, 1997) p. 221.

10. M. Legros, G. Dehm, R-M. Keller-Flaig, E. Arzt, K.J. Hemker, and S. Suresh, *Materials Science and Eng. A* (2001) in press.

11. G. Dehm and E. Arzt, *Applied Physics Letters* **77**, 1126 (2000).

12. G. Dehm, D. Weiss and E. Arzt, *Materials Science and Eng. A* (2001) in press.

13. B. Inkson, G. Dehm and T. Wagner in *Physical Sciences II* , edited by J. Gemperlova, I. Vavra (Proceedings of the 12th European Congress on Electron Microscopy, Brno, Czech Republic, 2000) p. 539.

14. B. Inkson, R. Spolenak and T. Wagner, EMAG Inst. Phys. Conf. Ser. **161**, 335 (1999).

15. R-M. Keller, S.P. Baker and E. Arzt, *J. Mater. Res.* **13**, 1307 (1998).

16. G.G. Stoney, *Proc. R. Soc. A* **82**, 172 (1909).

17. G. Dehm, F. Ernst, J. Mayer, G. Möbus, H. Müllejans, F. Phillipp, C. Scheu, M. Rühle, *Z. Metallkunde* **87**, 898 (1996).

18. A. Strecker, U. Salzberger, J. Mayer, *Prakt. Metallogr.* **30**, 482 (1993).

19. J. Marien, J.M. Plitzko, R. Spolenak, R-M. Keller, and J. Mayer, *J. of Micros.* **194**, 71 (1999).

20. B. Roos and F. Ernst, *J. Cryst. Growth* **137**, 457 (1994).

21. F. Ernst, R. Raj, and M. Rühle, *Z. Metallkunde* **12**, 961 (1999).

22. W. Mader, *Z. Metallkunde* **80**, 139 (1989).

23. M. Yu. Gutkin, M. Militzer, A. E. Romanov, and V. I. Vladimirov, *Phys. Stat. Sol. (a)* **113**, 337 (1989).

24. W.P. Vellinga, J.T.M. De Hosson, and V. Vitek, *Acta Mater.* **45**, 1525 (1997).

25. A. E. Romanov, T. Wagner, and M. Rühle, *Scripta Materialia* **38**, 869 (1998).

26. E. Arzt, G. Dehm, P. Gumbsch, O. Kraft, and D. Weiss, *Progress in Materials Science* **46**, 283 (2001).

27. B. von Blanckenhagen, P. Gumbsch, and E. Arzt, in this volume (2001).

Mat. Res. Soc. Symp. Proc. Vol. 673 © 2001 Materials Research Society

Observations of Dislocation Motion and Stress Inhomogeneities in a Thin Copper Film

T. John Balk, Gerhard Dehm and Eduard Arzt
Max-Planck-Institut für Metallforschung, Seestrasse 92, 70174 Stuttgart, Germany

ABSTRACT

In situ transmission electron microscopy has been utilized to study dislocation plasticity in a 200 nm thick copper film. The behavior of dislocations in a [111]-oriented grain was recorded during a thermal cycle. During cooling, it was observed that dislocations were emitted from a grain boundary triple junction in regular intervals of 30°C to 40°C. Subsequent glide occurred on a (111) plane parallel to the film surface, despite the expectation of zero resolved shear stress on such planes. The initial emitted dislocations remained close to the triple junction, avoiding contact with another [111] grain rotated by 17°. Glide into the opposite end of the grain was initiated only after the injection of several additional dislocations, which induced strong curvature in all dislocations near the active triple junction. *Post mortem* examination of dislocation curvature revealed that an inhomogeneous stress state existed within the grain.

INTRODUCTION

The mechanical behavior and reliability of thin metal films have been widely studied, but knowledge of the exact mechanisms that carry inelastic deformation is incomplete. Understanding the role of dislocations in thin film plasticity is critical for the testing and confirmation of existing models, e.g. the dependence of film strength on film thickness [1-3], and for the creation of new models.

Transmission electron microscope (TEM) observations of dislocation behavior have been made for several face-centered cubic metals, including copper [4], and indicate that dislocations tangle during cooling of the film, leading to high dislocation densities. These observed densities do not appear to increase as a result of repeated thermal cycling, indicating that deformation microstructure is alternately healed and regenerated during heating and cooling of the film. Allen *et al.* [5] performed *in situ* thermal cycling experiments in the TEM, using 200 nm thick Al films deposited on single crystalline Si coated with amorphous SiN$_x$. They observed a reversible and repeatable evolution (emission and recession) of dislocation loops, which alternately expanded and contracted within a <110> grain during cooling and heating, respectively. Such observations agree with the thermomechanical behavior of thermally cycled thin films, which exhibit highly repeatable stress-temperature curves.

It has been suggested that unpassivated metal films which do not form a native oxide may undergo constrained diffusional creep, whereby atoms diffuse from the grain boundaries to the free film surface in a film under tension at sufficiently high temperature [6]. The resulting diffusional "wedges" cause the film stresses to be partially relaxed at the grain boundaries, although the stress in the grain interior is not relaxed, assuming that the film does not slide relative to the substrate. This leads to an inhomogeneous stress state, with shear stresses acting on planes parallel to the film/substrate interface. There is no direct experimental evidence of this mechanism yet, although observations from the current study could be explained by this model.

EXPERIMENTAL DETAILS

Polycrystalline copper thin films, 200 nm in thickness, were deposited onto (100)-oriented Si substrates that had previously been coated with amorphous SiO$_x$ and an amorphous SiN$_x$ diffusion barrier. Deposition was performed at room temperature, using an ultrahigh vacuum (UHV) magnetron sputtering apparatus. Immediately following deposition, the films were annealed in the UHV chamber for 10 minutes at 600°C, and were allowed to fully cool to room temperature before removal. This heat treatment produced a columnar grain structure and a strongly [111]-textured film, as confirmed by θ-2θ x-ray scans, with practically no [100] grains present. Focused ion beam analysis revealed that the grains were equiaxed in the film plane and contained a large population of twins.

In order to determine their thermomechanical behavior, films were thermally cycled between 40ºC and 500ºC in a wafer curvature apparatus, at a rate of 6ºC/min. A maximum compressive stress of 260 MPa was achieved in the course of heating the 200 nm film to 500ºC, while a maximum tensile stress of 640 MPa developed during subsequent cooling to 40ºC.

Plan-view TEM specimens were made from a wafer that had previously been cycled in the wafer curvature apparatus. In the specimens, which are schematically depicted in figure 1, the copper film is completely intact, and only the Si substrate has been thinned by mechanical polishing. Although the Si substrate is removed in the electron transparent region of the disk, there exists a stable ring of Si around the dimpled region (figure 1), which provides a sufficiently rigid support for the copper film and forces it into compression/tension during heating/cooling.

In situ TEM thermal cycling experiments were performed at 200 kV in a JEOL 2000FX equipped with a double-tilt heating stage. Temperature was manually ramped using current control. The TEM cycles were performed by heating and cooling between 40ºC and 500ºC, in order to provide as direct a comparison with the wafer curvature tests as possible. The evolution of dislocation microstructure during thermal cycling was recorded at 25 frames per second using a TV rate camera and S-VHS VCR. For both bright-field and weak-beam TEM micrographs, a beam direction **B** near [111] and a two-beam diffraction condition of $\mathbf{g} = \bar{2}20$ were used. Weak-beam observations were recorded using a $(\mathbf{g}, 1.7\mathbf{g})$ condition.

EXPERIMENTAL RESULTS

Prior to thermal cycling and during heating

At the outset of an *in situ* heating cycle, it was observed that dislocations in a larger grain of the 200 nm Cu film were bowed out from the vicinity of a grain boundary triple junction. As shown in figure 2a, which is a single, unfiltered frame taken from the beginning of the video (T=25ºC), there are 5 dislocations in the grain, four of which (A-D) exhibit significant curvature away from the lower left triple junction. Figure 2b offers a schematic view of the 5 dislocations visible within the grain. During heating to 500ºC, dislocations A-D gradually contracted toward and disappeared into the triple junction, while the upper dislocation (E) underwent glide to the

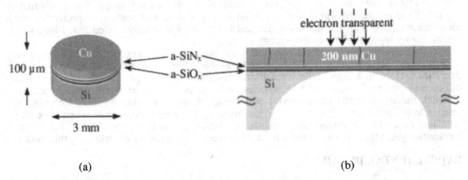

(a)   (b)

Figure 1. Schematic depiction of the plan-view TEM specimen that was subjected to *in situ* thermal cycling. (a) A disk 3 mm in diameter, bored from the Cu-coated Si wafer, was polished to approximately 100 µm in thickness by grinding away Si. (b) The thin disk was then dimpled and etched from the Si side to produce an electron transparent region in the center of the disk.

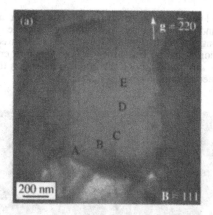 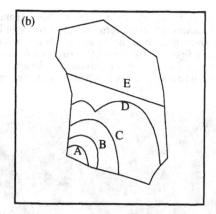

Figure 2. (a) TEM micrograph, taken from the *in situ* video at the outset of the thermal cycle, and (b) schematic of 5 dislocations (A-E) that are visible in the observed grain. Four of the dislocations (A-D) are bowed out from the grain boundary triple junction at the lower left. Note that dislocation D is pinned along its length.

grain boundary at the top of figure 2 and disappeared. At 500ºC, immediately prior to cooling, only one dislocation remained in the grain. This dislocation had been observed to result from a dislocation-dislocation reaction and remained sessile during cooling.

Cooling from 500ºC to 40ºC

During cooling to 40ºC, dislocations were sequentially emitted from the lower left triple junction of the grain in figure 2 and underwent glide parallel to the film plane. This behavior was recorded using weak-beam TEM and is presented in figure 3. Figure 3a shows the grain at 500ºC, at which point only one dislocation exists (the sessile reaction product mentioned previously). The sample was tilted through several diffraction conditions to confirm the absence of additional defects such as twins and other dislocations. During cooling, dislocations were emitted in succession, roughly every 30ºC to 40ºC, from the lower left triple junction. The first dislocation to be emitted remained in the leading position during the entire cooling segment, and none of the dislocations was passed by a subsequent dislocation.

At 355ºC, the first dislocation was emitted and immediately advanced into the grain; it is shown at a later time in figure 3b. It came to rest in the vicinity of the sessile dislocation. Whether the sessile dislocation acted as an obstacle to the motion of other dislocations is not completely clear, although it did not appear to do so. In figure 3b, dislocation #2 is about to be emitted, and one frame (40 msec) later in figure 3c, the second dislocation has advanced into the grain and pushed the first dislocation slightly forward. Figure 3d shows that, upon its emission, dislocation #3 is not able to advance as far into the grain as its predecessors. Note that dislocation #1 has encountered a pinning point, the same one that was recognized in figure 2. With the presence of a fourth dislocation in figure 3e, the leading dislocation has bowed even more around the pinning point. Note that, despite the high stress that must exist in order to bow the leading dislocation to this extent, all dislocations terminate on the lower grain boundary, avoiding the grain boundary on the right hand side. 5 frames (200 msec) later, in figure 3f, dislocation #5 enters, but is not able to push the first four dislocations very far and advances only a short distance itself. Some time later, in figure 3g, dislocation #8 has just been emitted; it is visible as a very small half-loop, bowing out from the lower left triple junction. Note that dislocation #3 is pinned at the same point as dislocation #1 in figures 3d-f. Almost two seconds later, in figure 3h, dislocation #8 has expanded and pushed forward a bit farther, but all dislocations still avoid terminating on the right hand grain boundary. Instead, they terminate on the lower grain boundaries and are therefore strongly bowed. Finally, four frames (160 msec) later in figure 3i, the cumulative stress on dislocation #1 has pushed it forward into the upper end of the grain. During this advance, dislocation #1 ran quickly along the grain boundary on the right and came to rest only when it reached the upper right grain boundary.

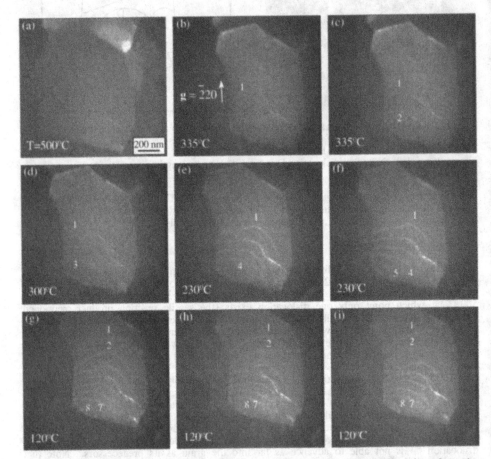

Figure 3. Sequence of video frames showing the sequential emission of dislocations from the lower left triple junction during cooling. The leading and recently emitted dislocations are numbered (in white) in each frame. Following their emission, additional dislocations are not able to advance as far as previous dislocations; compare figures 3c, 3f and 3g. A pinning point, located approximately in the center of the micrographs, acts on all dislocations as they pass (figures 3d-i). All dislocations avoid terminating on the right hand side grain boundary, instead bowing out strongly from the lower grain boundaries (figures 3g,h). Only under significant cumulative stress, induced by the later dislocations, does the first dislocation finally advance into the far end of the grain (figure 3i). See the text for further details.

Holding at 40°C after thermal cycling

During the final stages of cooling and while holding the specimen temperature at 40°C, a total of 6 dislocations advanced into the far end of the grain, leaving 4 dislocations bowing out from the triple junction, as shown in figure 4. In addition to the sessile dislocation, ten glissile dislocations are visible, with the tenth dislocation somewhat obscured by the bright contrast at the bottom. The dislocation density was calculated from the micrograph by measuring the total length of the 10 active dislocations and dividing by the grain volume, yielding $\rho \approx 4 \times 10^9$ cm^{-2}. This is in good agreement with other TEM studies of thermally cycled polycrystalline copper [4].

The fact that all dislocations avoided terminating on the right hand side grain boundary during cooling is most likely due to the misorientation of grains. The observed grain and all of its neighbors are [111]-oriented, with varying in-plane rotations between each grain pair. These misorientations were measured using selected area diffraction in the TEM, and the rotation angle of each neighboring grain relative to the observed grain is given in figure 5. It is seen that the right hand grain boundary (marked *RGB* in figure 5), which was avoided by all dislocations during cooling, borders a grain that is rotated by 17° relative to the central grain.

DISCUSSION

From the [111] orientation and dimensions of the grain, and from the observed distance of travel of the dislocations, it is clear that glide has occurred on a (111) plane parallel to the film surface. Such glide would not be expected in this thin film, which is under biaxial tension and should therefore have zero resolved shear stress on planes parallel to the film. Considering the dislocation behavior observed during cooling (figure 3), it is apparent that the dislocations were

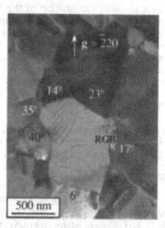

Figure 4. Weak-beam TEM micrograph of the observed grain, following complete cooling to room temperature. Note the significant curvature of dislocations bowed out from the triple junction, as opposed to the relatively straight dislocations in the far end of the grain, indicating that an inhomogeneous stress state exists across the grain.

Figure 5. Lower magnification, bright-field TEM micrograph of the central and neighboring grains, all of which were [111]-oriented. Superimposed on each neighboring grain is the misorientation relative to the central grain. All active dislocations avoided terminating on the right hand side grain boundary RGB, which is rotated by 17°.

pushed forward by subsequent dislocation emission. Additionally, considering the lack of any significant curvature in dislocations #1-6 at the top of figure 4, no shear stress appears to exist on the (111) plane in this end of the grain. Taken together, these observations indicate that a significantly inhomogeneous stress state exists within the grain.

Given the strong bowing of dislocations away from the triple junction, it appears that a significant stress acted to push the dislocations into the grain during cooling. The curvatures of dislocations near the triple junction emission point and those in the opposite end of the grain (upper portion of figure 4) were measured. The stress acting to bow out each dislocation can be estimated from the curvature using the following equation [7]:

$$\tau = \frac{\mu b}{R} \tag{1}$$

where μ is the shear modulus, b the Burgers vector (2.56 Å) and R the radius of curvature. The shear modulus for <110>{111} glide, $\mu = 40.8$ GPa, was calculated from the elastic constants for copper [8]. Starting with dislocation #10, closest to the triple junction, and moving up through dislocation #7 (figure 4), the approximate bowing stresses are 109, 60, 57 and 56 MPa. The 6 dislocations in the opposite end of the grain exhibit very little curvature, yielding a stress of less than 10 MPa in every case. It appears that the stress state in the grain is indeed inhomogeneous.

An inhomogeneous stress state could be caused by grain misorientations, but could also be due to partially relaxed stress at the grain boundaries. The observations in the current study thus provide possible support for the model of grain boundary diffusional relaxation, as suggested by Gao et al. [6]. It should be noted that, in a convergent beam electron diffraction study, Kramer et al. [9] also observed stress inhomogeneities across grains in Al which, due to the passivating effect of the native oxide, should not experience diffusional relaxation of the grain boundaries. But as a possible explanation of the behavior observed here, Gao's model [6] recognizes that diffusion of atoms from the grain boundaries to the bare, unpassivated copper surface could, at elevated temperatures, fully relax the stress between grains. However, the inner regions of grains would still carry the large biaxial stresses that exist in the film, leading to an inhomogeneous stress state as well as the presence of shear strains near the partially relaxed boundaries and parallel to the film/substrate interface. Such shear strains could explain the emission and glide of dislocations on a (111) plane parallel to the film, as was observed here.

CONCLUSIONS

This study of dislocation activity in a 200 nm thick Cu film during thermal cycling has revealed:
(1) Emission and unexpected glide of dislocations on a (111) plane parallel to the film surface.
(2) Stress inhomogeneities across a [111]-oriented grain, as indicated by dislocation curvature.

REFERENCES

1. L.B. Freund, J. Appl. Mech. **54**, 553 (1987).
2. W.D. Nix, Metall. Trans. **20A**, 2217 (1989).
3. W.D. Nix, Scripta Mater. **39** (4/5), 545 (1998).
4. G. Dehm and E. Arzt, Appl. Phys. Lett. **77** (8), 1126 (2000).
5. C.W. Allen, H. Schroeder and J.M. Hiller in *Thin Films – Stresses and Mechanical Properties VIII*, edited by R. Vinci, O. Kraft, N. Moody, P. Besser and E. Shaffer (Mater. Res. Soc. Proc. **594**, Warrendale, PA, 2000), pp. 123-128.
6. H. Gao, L. Zhang, W.D. Nix, C.V. Thompson and E. Arzt, Acta Mater. **47** (10), 2865 (1999).
7. D. Hull and D.J. Bacon, *Introduction to Dislocations*, 3rd ed. (Butterworth-Heinemann Ltd., Oxford, 1995) pp. 80-83.
8. J.P. Hirth and J. Lothe, *Theory of Dislocations*, 2nd ed. (Krieger Publishing Co., Malabar, Florida, 1992) p. 837.
9. S. Kramer, J. Mayer, C. Witt, A. Weickenmeier, M. Ruhle, Ultramicroscopy **81**, 245 (2000).

Dislocations and Deformation Mechanisms in Thin Films and Small Structures

Mat. Res. Soc. Symp. Proc. Vol. 673 © 2001 Materials Research Society

SOLID SOLUTION ALLOY EFFECTS ON MICROSTRUCTURE AND INDENTATION HARDNESS IN PT-RU THIN FILMS

Seungmin Hyun*, Oliver Kraft**, and Richard P. Vinci*
* Department of Materials Science and Engineering, Lehigh University, Bethlehem,
   Pennsylvania 18015
** Max Planck Institut fur Metallforschung, Stuttgart, Germany

ABSTRACT

The elastic moduli and flow stresses of as-deposited Pt and Pt-Ru solid solution thin films were investigated by the nanoindentation method. The influence of solid solution alloying was explored by depositing Pt-Ru solid solution thin films with various compositions onto Si substrates. The 200 nm films were prepared by DC magnetron cosputtering with a Ru composition range from 0 to 20wt%. As expected, the modulus and the flow stress both increased significantly with an increase in Ru. The experimental results compare favorably to predictions based on a simple dislocation motion model consisting of three strengthening terms: substrate constraint, grain size strengthening and solid solution strengthening.

INTRODUCTION

The general effects of solid solution alloying on microstructure development and strength are well known for bulk metals [1]. The increase in strength that is typically observed is the result of interaction between the stress field of a dislocation and the stress field associated with a solute atom. Strengthening mechanisms active in metal thin films have also been examined extensively [3,4]. However, the mechanical properties of solid solution alloy thin films have not been heavily studied because interest has traditionally focused on pure metals such as Al and Cu. As a result, the effectiveness of solid solution alloying relative to other strengthening mechanisms unique to thin films is not well understood.

In this work, thin films of Pt alloyed with as much as 20 wt% Ru were studied with the goal of improving our fundamental understanding of thin film strengthening behavior. The equilibrium binary phase diagram of the Ru and Pt system [2] shows a homogeneous solid solution up to 46 wt% Ru at 1000°C, so we remained well within the solubility limit. In bulk, small amounts of Ru in Pt dramatically increase the hardness of the alloy [1]. It is therefore reasonable to expect that large strength gains should be found in our films if solid solution alloying can play a significant role compared to the other strengthening mechanisms that are active.

EXPERIMENT

Pt-20wt%Ru and pure Pt targets were used for DC-magnetron cosputtering at ambient temperature on to uncooled substrates at a base pressure of 4.5×10^{-9} Torr. All substrates were 3 inch diameter Si (100) wafers with a thermally grown oxide layer approximately 100 nm thick. The substrates were pre-cleaned with an ion gun to improve the adhesion between film and substrate. The deposition rates for the two targets were very similar: approximately 25nm/min at

160 W sputter power and about 8.3nm/min at 50 W sputter power. Approximate film composition was controlled by independently controlling the sputter power for the two targets. After film fabrication, the thickness of the films was found to fall in the range of 200-240nm, as measured by AFM. Certain samples were annealed at 800°C in a N_2 atmosphere to observe microstructure changes. TEM samples were prepared by chemically etching the substrate from the wafer backside with a solution of HF, HNO_3, and CH_3HCOOH, thereby creating windows of the Pt and Pt-Ru films. The back-etched samples were mounted on copper grids using M-Bond adhesive. A Precise Ion Polishing System (PIPS) was used to further thin the films so that they could be imaged using a 200kV JOEL 2000 FX microscope. A tracing technique was performed to measure the average grain sizes of the films. The hardness of the as-deposited Pt and Pt-Ru films at room temperature was measured by the nanoindentation method (Hysitron TriboScope on a Digital Instruments small-sample AFM). A Berkovich diamond tip was used for the indentations, all of which were performed at a 200 μN maximum load. The indented areas of the films were observed by scanning with the indenter tip to confirm that there was no pile-up.

RESULTS & DISCUSSION

Microstructure

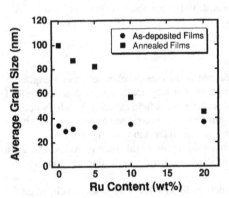

Figure 1. Average Grain Size of Films

The microstructure of the as-deposited and annealed films was examined to observe the extent of any solid solution alloy effect on grain size. Figure 1 shows the measured average grain sizes of the films. The grain sizes of the as-deposited films, which are around 33 nm, are not dependent on Ru content. However, the grain sizes of the annealed films decrease with an increase in the amount of Ru. The average grain size of the pure Pt film is 100nm, while that of a Pt-10% Ru film is only 56nm, as shown in Figure 1. This behavior can most likely be explained by the drag effect of solute atoms on grain boundary motion. However, from these observations it cannot be determined if the measured grain sizes are the maximum possible for each composition or if they simply indicate significant differences in growth rate.

Nanoindentation

Hardness and modulus of the as-deposited films, determined from the load-displacement data, are shown in Figure 2. The values were obtained by the Oliver-Pharr method [5]. The hardness of the Pt film and the Pt-20wt%Ru film are 5.1 GPa and 8.5 GPa, respectively. Representative AFM images of indented areas of the films are shown in Figure 3. The indented area of the Pt film is clearly larger than that of the Pt-20wt%Ru film, in agreement with the numerical results. These results are discussed in more detail in the context of the model calculations below.

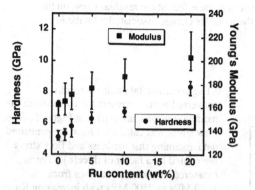

Figure 2. Hardness and Modulus of as-deposited Films

Figure 4 shows calculated and measured values of the moduli. The calculated moduli were obtained from the Hashin and Shtrikman equation [6], which assumes that the films are isotropic and homogeneous composite materials. This assumption is not completely valid for a film with some degree of texture, but serves as a reasonable starting point for comparison. As expected, the moduli increase with an increase of Ru content: Young's modulus of the pure Pt film is 164 GPa, and the modulus increases with Ru content up to 204 GPa in the

20% Ru film($v \approx 0.39 \sim 0.3658$ according to Ru content). The calculated Young's modulus is mostly in the error bar range of the measured modulus. The modulus of the Pt film was similar to that of bulk Pt (171 GPa).

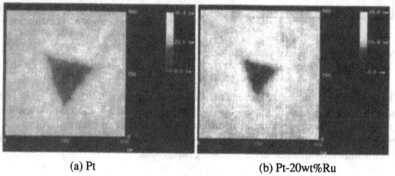

(a) Pt   (b) Pt-20wt%Ru

Figure 3. AFM Images of Indented Areas of as-deposited Films

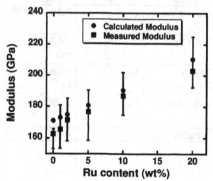

Figure 4. Measured and Calculated Young's Modulus of as-deposited Films

Bahr et al. [7] have investigated the moduli of Pt films prepared by evaporation and RF-sputtering. They found that their evaporated Pt film had a low modulus in the range of 120-140GPa, but their sputtered Pt film showed a higher modulus, 205 GPa. The low modulus of evaporated Pt was explained as possibly resulting from the tensile residual stress in the films. However, the high modulus associated with the sputtered Pt films did not fully support this argument. In our study, compressive residual stresses were measured in the as-deposited films using the wafer curvature method. The residual compressive stress increased with an increase in Ru. If we assume that the effect of the residual stress on the modulus measurement can be ignored, then the modulus change associated with the Ru represents a true solid solution alloy effect.

Flow Stress Calculations

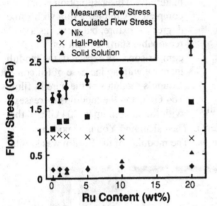

Figure 5. Measured and Calculated Flow Stresses of Films

The experimental hardness results, converted to flow stresses for comparison to modeling results, are shown in Figure 5. The flow stress was calculated from the measured data assuming that hardness and flow stress are related by a factor of three[8]. The measured flow stress increases from 1690 MPa to 2800 MPa with increasing Ru content, which shows that solid solution alloying can have a significant effect on the strength of thin films. Calculated flow stresses, also shown in Figure 5, were derived using common models based on dislocation motion. The final flow stress calculation is summarized by equation (1):

$$\sigma_f = \sigma_{interface} + \sigma_{gs} + \sigma_{ss} \qquad \text{eq.(1)}$$

The three strengthening terms are the contributions of substrate constraint ($\sigma_{interface}$), grain size strengthening (σ_{gs}) and solid solution strengthening (σ_{ss}). As seen in this equation, we assume that these three mechanisms can act simultaneously to impede dislocation motion. All parameters and relations used in the calculations are shown in Table 1, and the models associated with each mechanism are explained in the following paragraphs.

Table 1. The parameters and relations for the flow stress calculation.

Parameter	Symbol	Value	Parameter	Symbol	Value
Burger's vector	b	$2.77 \sim 2.753 \times 10^{-10}$m	Poisson'ratio	ν	$0.39 \sim 0.3658$
Shear modulus	μ_f Film μ_s Substrate	$60.6 \sim 75.77$GPa 66.5GPa	Grain Size	d	$29.2 \sim 36.5$ nm
Constant	B βs	$0.9025 \sim 0.9085$ $0.829 \sim 0.834$	Hall-Petch Constant	k	0.18 MPa m$^{1/2}$
	$\dfrac{\sin \phi}{\cos \phi \cos \lambda}$	3.464	Line tension	E_L	$1/2\mu b^2$
Thickness	h	$200 \sim 240$nm	Constant	A	0.5
Average solute atom distance from a slip plane	z	$\dfrac{b}{\sqrt{6}}$			

Obstacle strength	$K \approx \dfrac{2E_L}{10} \approx \mu b^2 \sqrt{\delta^2 + \eta^2 \beta^2}, \beta \approx \dfrac{1}{20}$
	$\eta \approx \dfrac{1}{\mu}(\dfrac{d\mu}{dc})$, c : Solute atom content
	Misfit parameter, $\delta \approx \dfrac{1}{a}\dfrac{da}{dc}$, a : Lattice parameter

The strengthening of thin films occurs, in part, because dislocation motion is constrained by the substrate and inhibited by grain boundaries. The substrate-strengthening model we used was developed by Nix [9], following the approach introduced by Freund [10]. The equation describing the stress is

$$\sigma_{interface} = \frac{\sin\phi}{\cos\phi\cos\lambda} B \frac{b}{2\pi(1-v)h} [\frac{\mu_f \mu_s}{\mu_f + \mu_s} \ln(\frac{\beta_s h}{b}) + \frac{\mu_f \mu_o}{\mu_f + \mu_o} \ln(\frac{\beta_o t}{b})] \qquad \text{eq.(2)}$$

where μ_f, μ_s and μ_o are the shear moduli of the film, substrate, and oxide, h and t are the thickness of the film and oxide, and β_s and β_o are constants, and ϕ and λ are the included angles between the glide plane normal direction and Burgers vector and the film normal direction, respectively, as described in reference [9]. Because Pt has no intrinsic passivation layer, the oxide strengthening term that is shown above equation is not considered in this calculation. The only components of this model that are expected to change with increasing Ru content are the shear modulus and Burgers vector. No obstacles to dislocation motion or grain size strengthening are considered.

The grain boundary contribution to strengthening may be described by the Hall-Petch relation:

$$\sigma_{gs} = kd^{-\frac{1}{2}} \qquad \text{eq.(3)}$$

where d is the grain diameter and k is the Hall-Petch coefficient. To the best of our knowledge, no Hall-Petch coefficient has been determined experimentally for Pt, so our value was obtained via calculation [11].

The flow stress of a pure metal thin film is frequently described as simply the sum of the dislocation constraint and grain size strengthening mechanisms. Work hardening is often added to account for the discrepancy between measured film strengths and the results of the simple models. However, in our calculations a solid solution hardening mechanism must also be taken into account. The model for this contribution is based on the strain field in the Pt matrix created by the distribution of solute atoms which blocks dislocation motion. A well established equation for solid solution strengthening is:

$$\sigma_{ss} = A\frac{K_{max}^{4/3}c^{2/3}z^{1/3}}{E_L^{\frac{1}{3}}} \qquad \text{eq.(4)}$$

where c is the concentration of solute atoms, E_L is a line tension, z is average distance between solute atom and slip plane, K is an obstacle strength and A is constant that is adjusted for this equation. This model was originally designed by Labusch [12]. Our version of the equation, along with its parameters, is simplified for FCC solid solution strengthening following Haasen [13].

As seen in Figure 5, the calculated flow stress increases from 1100 MPa to 1620 MPa with the increase in Ru content. The agreement with the measured results is reasonable, considering the simplicity of the modeling approach. The calculated flow stresses and experimentally measured flow stresses of Pt-Ru thin films show the solid solution hardening effect. It should be noted that the interface constraint mechanism and grain size strengthening mechanism do not show large changes with an increase of Ru because the shear modulus, Burgers vector and grain size do not vary much in the as-deposited films. In contrast, the solid solution strengthening model predicts a significant increase in strength similar to the increase that was measured. This result supports the conclusion that solid solution alloying can significantly increase the strength of thin metal films over a wide range of solute content. It is possible that the solute may also affect work hardening, but this potential contribution cannot be easily evaluated.

CONCLUSIONS

The hardness of the solid solution alloy films is increased by an increase in Ru content. The predicted flow stress and measured flow stress both increase and show similar tendencies. A flow stress model based on dislocation motion in thin films, which includes three strengthening terms due to substrate constraint, grain size strengthening and solid solution strengthening, does a reasonable job of predicting the flow stress change according to the Ru content. A comparison of the relative contributions of the three mechanisms considered reveals that the solid solution strengthening mechanism is the main mechanism associated with the film hardening.

REFERENCES

1. Metals Handbook, 'Properties and Selection ; Nonferrous alloys and special purpose materials, **2**, 707 (1990).
2. Binary Alloy Phase Diagrams, ed. T.B.Massalski, 2nd edition, ASM International (1992).
3. R.P.Vinci, E.M.Zielinski, and J.C.Bravman, Mat.Res.Soc.Symc.Proc., **356**, 459 (1995).
4. R.Venkatraman and J.C Bravman, J.Mater.Res., **7**, 2040 (1992).
5. W.C. Oliver and G.M.Pharr, J.Mater.Res., **7**, 1564 (1992).
6. G.Grimvall, Thermophysical Properties of Materials, North-Holland (1986).
7. D.F.Bahr, D.A. Crowson, J.S. Robach and W.W. Gerberich, Mat. Res. Soc. Symp. Proc. **505**, 85 (1997).
8. D.Tabor, The Hardness of Metals, Oxford at the Clarendon Press (1951)
9. W.D.Nix, Metal. Trans.A. **20A**, 2217 (1989).
10. L.B.Freund, J.Appl. Mech. **54**, 553 (1987).
11. J.G. Sevillano, Materials Science and Technology, Vol6, ed. H.Mughrabi, VCH (1993).
12. R.Labusch, Phys., Stat., Sol., **41**, 659 (1970)
13. Peter Haasen, Physical Metallurgy, Cambridge Univ.Press (1996)

ACKNOWLEDGMENTS

The work was supported by NSF DMR-0072134, subcontract PY-0826

Mat. Res. Soc. Symp. Proc. Vol. 673 © 2001 Materials Research Society

Lack of hardening effect in TiN/NbN multilayers

Jon M. Molina-Aldareguia, Stephen J. Lloyd, Zoe H. Barber and William J. Clegg
Dept of Materials Science and Metallurgy, University of Cambridge, CB2 3QZ, UK

ABSTRACT

There is evidence indicating that multilayer films can be harder than monolithic ones. To investigate this, TiN/NbN multilayers with bilayer thicknesses ranging from 4 nm to 30 nm have been grown on MgO (001) single crystals using reactive magnetron sputtering. The sharpness of the interface and the composition modulation, which would be expected to strongly influence dislocation motion, have been studied by X-ray diffraction (XRD). These experiments show that the interfaces remain reasonably sharp (interface thickness ~1 nm) and the composition modulation amplitude is maximum for multilayers with bilayer thicknesses greater than ~10 nm. With thinner bilayers, the composition modulation decreases but the layered structure remains. Despite this, the nanoindentation hardness of the multilayers is between 20 and 25 GPa, which is similar to that of TiN and NbN alone, and therefore, no hardening due to the layering is observed. The deformation mechanisms observed under the indent in the TEM are consistent with these results.

INTRODUCTION

It has been shown that the hardness of nitride multilayers can be greater than that of monolithic films of the components [1]. These results have been attributed to hardening due to the elastic mismatch between the layers [2], according to Koehler's ideas [3]. However, discrepancies still exist between the predictions and the experimental results. For instance, this paper examines the nanoindentation behaviour of TiN/NbN multilayers that do not show the hardening effect, although they posses a strong composition modulation.

EXPERIMENTAL

Multilayers were deposited using ultra-high vacuum reactive d. c. magnetron sputtering in an Ar-40% N_2 gas mixture on (001) oriented MgO single-crystals heated to 800 °C. Previous work [4] showed that the highest quality films were produced under conditions where the residual intrinsic stresses, due to bombardment by energetic species during deposition, are small and slightly compressive. Under deposition conditions that prevented bombardment of the films, open porous structures were observed. Where the induced compressive stresses where very high, it was difficult to maintain epitaxy in the films. In the current work, the multilayers were deposited at a sputtering gas pressure of 2.3 Pa and a substrate to target distance of 30 mm, to promote bombardment by energetic species during deposition. It was found that deposition rates of 0.35 nm s^{-1} for NbN and 0.05 nm s^{-1} for TiN allowed the deposition of 1 micron thickness epitaxial multilayers with different bilayer thicknesses, ranging from 4 nm to 25 nm. Under these conditions, the intrinsic stresses in monolithic TiN films were found to be slightly compressive (~1.6 GPa) and no stresses were found in NbN films.

The hardness and elastic modulus of the films for a range of indentation depths ranging from 50 nm to 400 nm were determined by nanoindentation (Micro Materials Ltd, NanoTest 600), using the analysis due to Oliver and Pharr [5]. X-ray diffraction (XRD) was carried out using a Philips X'Pert powder diffractometer with Cu K$_\alpha$ radiation. The microstructure of the as-deposited films and the deformed regions around the indents have been studied using transmission electron microscopy (TEM). TEM specimens, including cross-sections of the indents, were prepared in a focused ion beam (FIB) FEI FIB200 workstation using a technique described elsewhere [6], allowing large areas (typically 20 μm x 4 μm) of uniform thickness, suitable for TEM examination, to be produced. TEM was performed on a Philips CM30 microscope operating at 300 kV.

RESULTS

Characterisation of the multilayers

It is predicted that the sharpness of the interfaces and the intermixing between the layers should strongly influence dislocation motion [7]. Figure 1 shows XRD patterns from four different multilayers. For large bilayer thicknesses, two broad peaks, corresponding to the (002) planes of TiN and NbN can be seen (bilayer thickness Λ ~ 26 nm, in figure 1). As the bilayer thickness decreases and becomes comparable with the wavelength of the X-rays, the XRD patterns consist of a main Bragg peak at the position expected for an homogeneous film of the same composition and satellite peaks, whose spacing is determined by the bilayer thickness Λ. The composition modulation amplitude and the interface thickness influence strongly the decay in intensity of the satellite peaks (for more detailed explanations of XRD in multilayer structures, see McWhan [8]). The presence of high order satellite peaks in the XRD patterns of figure 1 is a good qualitative indication that the interfaces remain reasonably sharp and that the composition modulation is strong. Fitting of the experimental patterns in figure 1 by a specially developed model [9], indicates that interface thickness is ~ 1 nm and that the composition modulation is maximum for bilayer thicknesses greater than ~ 10 nm. For thinner bilayers, the composition modulation is reduced but the interfaces remain sharp (~ 1 nm).

Nanoindentation behaviour

Figure 2 shows two indentation loading-unloading curves for a ~ 13 nm bilayer thickness multilayer at two maximum indentation depths. Figure 3 plots the measured hardness and elastic modulus against indentation depth. They both increase as the indentation depth decreases, until a plateau is reached below an indentation depth of ~ 200 nm. The drop in hardness and elastic modulus for indentation depths above 200 nm was accompanied by a sudden pop-in event during loading of the indent (see the indentation loading-unloading curve *a* in figure 2). For very low indentation depths (below ~ 50 nm), both hardness and elastic modulus decrease slightly, probably due to roughness of the films or bluntness of the indenter tip. The hardness and elastic modulus of the mutilayers was determined, therefore, at indentation depths below 200 nm. The results are plotted against bilayer thickness in figure 4. They all lie between the measured values of TiN and NbN monolithic films deposited at the same conditions and show little effect of the layered structure.

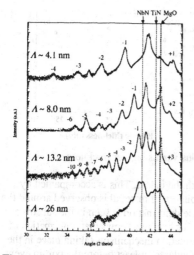

Figure 1. Showing XRD patterns of four of the multilayers under study. Bilayer thickness Λ is indicated in the figure.

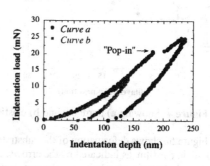

Figure 2. Showing two loading unloading curves for a 13 nm bilayer thickness multilayer, above and below the pop-in.

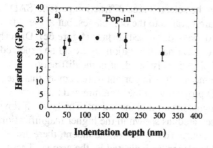

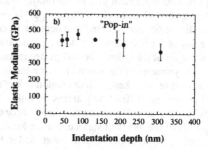

Figure 3. Showing measured **a)** hardness and **b)** elastic modulus of a 13.1 nm bilayer thickness multilayer against indentation depth.

TEM observations

The pop-in behaviour described in the previous section was characteristic of all the nitride films and multilayers deposited. To study the origin of the pop-in, indents of different sizes were studied in the TEM. For instance, figure 5a and 5b show cross-sectional TEM images of two indents, one for indentation depths above the pop-in event mentioned above (curve *a* in figure 2) and the other, below the pop-in event (curve *b* in figure 2). These micrographs were taken from a multilayer deposited with an electrical bias voltage of 300 V at the substrate.

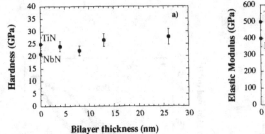 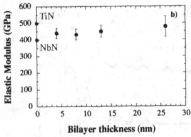

Figure 4. Showing **a)** hardness and **b)** elastic modulus versus bilayer thickness.

Figure 5a shows deformation of the substrate beneath the indent. This is accompanied by cracking in the film, as indicated by the arrows, and no pile-up of material is observed around the indent. At indentation depths below the pop-in event, little deformation is seen in the substrate and the material piles-up around the indent (figure 5b). This example shows that deformation of the substrate beneath the indent affects strongly the deformation mechanisms taking place in the multilayer, and illustrates the need to study very shallow indents, at least before the pop-in event during loading, is spite of the greater difficulty in locating and sectioning them for TEM analysis.

The deformation process under the indent is very complex. Figure 6a show a cross-sectional image of a 20 mN indent in a 13 nm bilayer thickness multilayer, taken close to the 100 Z.A with the layers parallel to the electron beam, as shown in the selected area diffraction pattern (SAD DP) of figure 6b. Away from the indent the layers are parallel to the 002 planes, but at the sides of the indent the 002 planes are no longer parallel to the undeformed surface (figure 6c). Directly below the tip of the indent, the 002 planes retain their original orientation, as would be expected from the symmetry of the indenter. The rotation of planes is also evident in the diffraction pattern (figure 6d) taken from the region below the indent. It is important to remember that the TEM images and diffraction patterns are only a 2 D projection of the structure and that the rotation of planes will be occurring in the three dimensions around the indent. The layers follow the profile of the indented surface, but are locally serrated, as shown in the higher magnification image of figure 6e (schematically illustrated in figure 6f). To the side of the indent, there are regions of localised deformation running along (011) planes, as indicated by the arrows. These planes agree with the expected slip system {110}<1$\bar{1}$0> for TiN [10]. The shearing along these planes causes discontinuities in the layering evident in figure 6 e.

Figure 5 a) and **b)** are cross-sectional TEM images of indents at maximum loads of 25 mN and 15 mN, respectively. They correspond to indents above and below the pop-in event shown in figure 2.

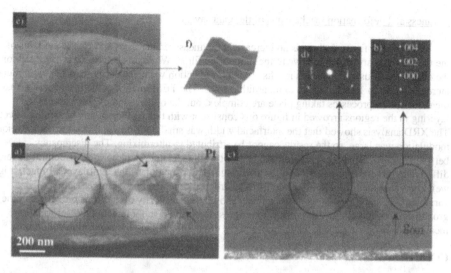

Figure 6. 20 mN indent in a TiN/NbN 13.1 nm bilayer thickness multilayer image down the (001) Z.A. **a)** BF image **b)** DP away from indent **c)** 002 DF image and **d)** DP beneath the indent. The arrows show localised deformation along (011) planes, which breaks the layer through.

DISCUSSION

Substrate effects in the measured hardness

The nanoindentation study of the multilayers displays behaviour that is characteristic of hard thin films deposited on soft substrates (hardness and elastic modulus of MgO are 9 GPa and 300 GPa, respectively). The substrate affects the measurements above certain critical indentation depth. In the present case, the loading unloading curves show a pop-in event at indentation depths of approximately 200 nm. TEM observations of indents before and after the pop-in (figure 5) show that this is due to sudden deformation of the substrate below the indent. At this point, the contact depth is around 140 nm, or 0.14 of the total thickness. Below this indentation depth, the measured hardness is constant down to very shallow indents (figure 3a) and does not seem affected by the substrate. This is in good agreement with the general rule that to obtain a true film hardness the indentation depth must not exceed 10 % of the film thickness. Substrate influence is, however, far more significant in the determination of the elastic properties of the film and these are probably influenced by the substrate even below the pop-in (figure 3b). It was shown in figure 5a that substrate deformation was accompanied by cracking in the film. It is not clear whether cracking arises during loading or unloading. However, there are no traces of radial/median cracks in the surface around the indent or just beneath the indent (which are usually the first to form during loading). We suggest that the observed cracks are lateral cracks that form during unloading, aided by the elastic strain energy produced during unloading by the mismatch resulting from the permanent deformation in the substrate and the mismatch in the elastic properties of film and substrate.

Hardness and deformation mechanisms of the multilayers

As shown in figure 4, there is no increase in the measured hardness and elastic modulus of the multilayered structures relative to the monolithic films. While this behaviour is expected for the elastic modulus, the hardness results are in contradiction with results of others [1], who found increases of up to 100 % relative to monolithic films. The TEM images of the indents show that the deformation processes taking place are complex, but the observation of *shear* across the layering in the regions arrowed in figure 6 is consistent with the absence of any hardening effect. The XRD analysis showed that the interfacial width was small (~ 1 nm) and that the composition modulation was large, so the results cannot be attributed to intermixing. The discrepancy between this work and previous studies [1] is not well understood, unless some structural factor, different to the ones considered here, is affecting the deformation of the multilayers. In fact, it is well know that the hardness of nitride films is strongly dependent on the processing variables [11] and previous work has already shown the sensitivity of the films properties to the growth conditions [4]. Further work is in progress to establish in more detail the deformation mechanisms occurring under the indents and the effect of changing the deposition conditions.

CONCLUSIONS

TiN/NbN multilayers, with bilayer thicknesses ranging from 4 nm to 26 nm have been deposited by UHV d.c. reactive magnetron sputtering. XRD patterns show a well defined composition modulation. However, the hardness of the multilayers is comparable to the hardness of TiN and NbN alone, showing there is no hardening due to the presence of the interfaces. Moreover, TEM studies support this conclusion showing shearing across the layers in the regions surrounding the indents.

ACKNOWLEDGMENTS

Jon Molina Aldareguia is funded by the Basque Government Education Department through the Researchers Development Program.

REFERENCES

1. M. Shinn, L. Hultman and S. A. Barnett, *J. Mater. Res.* **7**, 901 (1992)
2. X. Chu and S.A. Barnett, *J. Appl. Phys* **77**, 4403 (1995)
3. J. S. Koehler, *Phys. Rev.* **B2**, 547 (1970)
4. J.M. Molina-Aldareguia, S. J. Lloyd, Z. H. Barber, M. G. Blamire and W.J. Clegg in *Thin Films-Stresses and Mechanical properties*, ed. by R. Vinci, O. Kraft, N. Moody, P. Besser and E. Shaffer II, (Mater. Res. Symp. Proc. **594**, Boston, MA, 2000) pp 9-15
5. W. C. Oliver and G.M. Pharr, *J. Mater. Res.* **7**, 1564 (1992)
6. S. J. Lloyd, J. E. Pitchford, J.M. Molina-Aldareguia, Z. H. Barber, M. G. Blamire and W. J. Clegg, *Microsc. Microanal.* **5** (Suppl 2), 776 (1999)
7. J. E. Krzanowski, *Scr. Met.* **25**, 1465 (1991)
8. D. B. McWhan in Syntethic Modulated Structures, ed. by L. L. Chang and B. C. Giessen (Academic Press, Inc., Orlando, 1985) pp. 43-74
9. J.M. Molina-Aldareguia and W.J. Clegg, not published
10. M. Odén, H. Ljungcrantz and L. Hultman, *J. Mater. Res.* **12**, 2134 (1997)
11. J.-E. Sundgren, *Thin Solid Films* **128**, 21 (1985)

Mat. Res. Soc. Symp. Proc. Vol. 673 © 2001 Materials Research Society

Temperature and Strain Rate Dependence of Deformation-Induced Point Defect Cluster Formation in Metal Thin Foils

K. Yasunaga, Y. Matsukawa, M. Komatsu and M. Kiritani
Academic Frontier Research Center for Ultra-high Speed Plastic Deformation
Hiroshima Institute of Technology, Miyake 2-1-1, Saeki-ku, Hiroshima 731-5193, Japan

ABSTRACT

The mechanism of plastic deformation in thin metal foils without involving dislocations was examined by investigating the variations in vacancy cluster formation during deformation for a range of deformation speeds and temperatures. The deformation morphology was not seen to change appreciably over a very wide range of strain rate, 10^{-4}/s – 10^{6}/s, whereas the number density of vacancy clusters was observed to increase with increasing strain rate up to saturation value that is dependent on materials and temperature. The density of vacancy clusters decreased to zero with decreasing deformation speed. The strain rate at which the density of vacancy clusters begins to decrease was found to be proportional to the vacancy mobility, suggesting that the vacancies are generated as dispersed vacancies and escape to the specimen surfaces during slow deformation without forming clusters. A very long tail in the distribution of the density of vacancy clusters towards lower strain rates was reasonably attributed to the generation of small vacancy complexes due to deformation. These results give valuable information that can be used to establish new models for plastic deformation of crystalline metals without involving dislocations.

INTRODUCTION

Through their previous research, the present authors have demonstrated that the plastic deformation of metal thin foils proceeds without crystal dislocations [1, 2]. As a consequence of heavy plastic deformation, which ends in fracture, vacancy clusters are produced at high density, in the form of stacking fault tetrahedra, even in aluminum.

Understanding the mechanism of vacancy cluster formation is believed to lead to the development of a new deformation mechanism model that does not involve dislocations. The experiments in this study were conducted to pursue variations in vacancy cluster formation according to deformation speed and deformation temperature. These experiments clarified the previously unresolvable question of whether vacancy clusters are directly formed by deformation or whether they are formed by the aggregation of dispersed vacancies introduced by deformation.

EXPERIMENTAL PROCEDURE

Annealed metal films (Al, Au, Cu and Ni, all 99.99% purity) in ribbon form (20 μm x 3 mm x 10 mm) were cut half way across in the middle of their length in order to initiate deformation at that point.

The elongation speed of the films varied over a wide range: 10^{-9} – 1 m/s. High-speed deformation was performed by pulling the ribbon apart at a speed of 1 m/s, and slow deformation was performed using a deformation rig, the cross-head speed of which could be set as low as 1 mm/week by gearing-down the drive motor in many stages. The deformation presently of interest

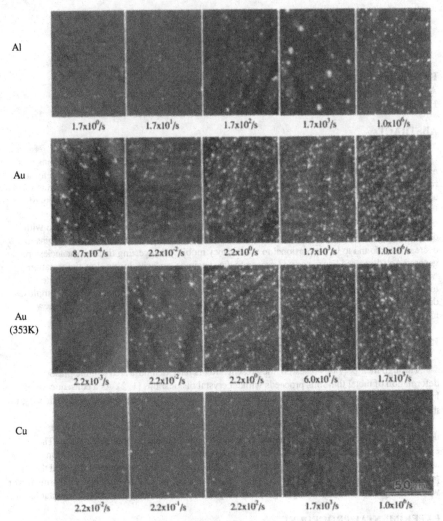

Figure 1. Variation of vacancy cluster formation with local strain rate in four kinds of metal foils deformed to fracture. Deformation at room temperature if not indicated otherwise.

proceeds in a very narrow band about 1 μm wide, and the deformation speed in terms of strain rate in the two extreme cases mentioned here are 10^6/s and 10^{-4}/s, respectively. All strain rates mentioned in this paper are those for this very narrow band.

Deformation at elevated temperatures was performed in heated Ar gas atmosphere. Annealing after deformation was performed on the heating stage of an electron microscope.

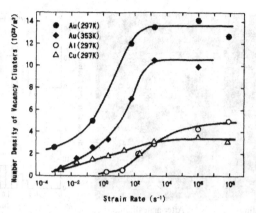

Figure 2. Variation of number densities of vacancy clusters with deformation speed.

All microstructure observations were performed under weak-beam dark field conditions in which the [200] reflection was excited. The acceleration voltage of the microscope was 200 kV in all cases except for Al, for which the voltage was dropped to 120 kV to prevent displacement damage.

EXPERIMENTAL RESULTS

Deformation Speed Dependence: Figure 1 shows the variation in vacancy cluster formation with strain rate for four metals. A common characteristic of the variation is the saturation of the number density at high strain rates, and the gradual decrease to zero at lower strain rates.

Figure 2 shows the variation of the number density of vacancy clusters with strain rate. All the variations have a similar character, with the saturation level differing among the different metals. Another notable feature is that the number density of vacancy clusters changes continuously over a very broad range of strain rates.

Deformation Temperature Dependence: The results for the deformation of Au at elevated temperature are also shown in Fig. 2. The saturation level seems to be similar to the case of lower temperature deformation, however the curve is shifted to higher strain rates, requiring higher strain rates to produce a given vacancy cluster density.

Annealing after Deformation: Vacancy clusters in Ni formed by deformation at room temperature are very small and grow during annealing at vacancy mobile temperatures [3] as shown in Figs. 3 and 4. Ni was annealed in this study because Ni lacks thermally activated vacancy motion at room temperature, which differs from the other three fcc metals.

DISCUSSIONS

Vacancy Cluster Formation from Deformation-Induced Dispersed Vacancies: It was demonstrated in a preceding paper in this volume based on various observations on vacancy cluster

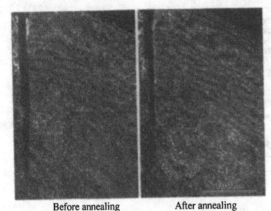
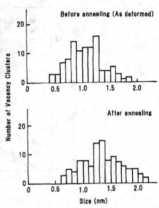

Before annealing   After annealing

Figure 3. Growth of vacancy clusters in Ni by post-deformation annealing.

Figure 4. Change of size distribution of vacancy clusters by annealing.

formation [2] that vacancy clusters are formed from dispersed vacancies generated by plastic deformation, and are not directly formed by deformation. Considerations on the variation with strain rate and temperature in this paper confirm this conclusion.

The general characteristics of the variation of vacancy cluster density with strain rate in three metals examined are that the number of vacancy clusters saturates at high strain rates, and that the vacancy cluster density decreases to zero at lower strain rates. However, no differences were observed in the deformation morphology of the deformed samples, such as the shape and thickness of fractured edge, even at the two extremes over 12 orders of magnitude of local strain rate. It is natural to understand the deformation mode to be the same, and vacancy type defects to be generated regardless of high and low strain rate. The difference in the vacancy cluster formation comes from the difference of reactions of vacancies after their generation.

Escape of Deformation-Induced Vacancies to Specimen Surfaces During Deformation: The time required to achieve a given level of deformation is inversely proportional to the strain rate. As such, the decrease of vacancy cluster density at small strain rates is attributable to the escape of vacancies to the sample surfaces during prolonged deformation. A very simplified analysis is given below.

The accumulation rate of vacancy defect concentration C during deformation is given as the difference between the defect generation rate by deformation and the escape of defects to the specimen surfaces, expressed as

$$dC/dt = G - C_s MC, \tag{1}$$

where G is the generation rate which is proportional to the strain rate ε_{rate}, $G = \alpha\varepsilon_{rate}$, C_s is the sink concentration, equal to approximately $(b/h)^2$, where b is the jump distance and h is the specimen thickness, M is the mobility of vacancy defects in unit of jumps/s. The accumulation of vacancies is then given by

$$C = (G/C_s M)[1 - \exp(-C_s Mt)]. \tag{2}$$

Table 1. Relation between saturation of vacancy cluster formation and vacancy migration rate (at room temperature when not indicated).

Material	Al	Au		Cu
$(C_s \varepsilon M_V)_{exp}$(s⁻¹)	5x10⁴	2.5x10¹	3x10²(353K)	7x10²
E_{1V}^m(eV)	0.65	0.83		0.72
M_{1V}(s⁻¹)	9.3x10¹	8.2x10⁻²	1.4x10¹	6.1x10⁰
$(C_s \varepsilon M_V)_{exp}/M_{1V}$	540	300	20	120
E_{2V}^m(eV)	0.50	0.74		0.64
M_{2V}(s⁻¹)	3.3x10⁴	2.8x10⁰	2.7x10²	1.4x10²
$(C_s \varepsilon M_V)_{exp}/M_{2V}$	1.5	9.0	1.1	5.1

E_{1V}^m and E_{2V}^m: Migration activation energy of single and di-vacancy, respectively. M_{1V} and M_{2V}: Jump rate of single and di-vacancy, respectively.

When the degree of deformation ε until fracture is postulated to be the same for different strain rates, the time required for the deformation is inversely proportional to the strain rate ε_{rate}, $t = \varepsilon/\varepsilon_{rate}$. The density of vacancy defects remaining after deformation is given by

$$C = (\alpha \varepsilon_{rate}/C_s M)[1 - \exp(-C_s \varepsilon M /\varepsilon_{rate})]. \qquad (3)$$

The values of $(C_s \varepsilon M)$ in eq. (3) were obtained by fitting the equation to the experimentally obtained strain rate dependence in Fig. 2, and are listed as $(C_s \varepsilon M)_{exp}$ in Table 1 for the four cases. The fitting here was made for the part at which the cluster density increased almost to the saturation levels (the lower part will be analyzed in the next section).

The jump rates of a single vacancy M_{1V} calculated with the known value of the migration activation energy E_{1V}^m [4] ($M_{1V} = \nu \exp(-E_{1V}^m/kT)$), are listed in the table, and the ratios $(C_s \varepsilon M)_{exp}/M_{1V}$ are also listed. This ratio should be similar among the four cases if the deformation modes are similar and the majority of vacancies occur as single vacancies. However, there is a difference of more than one order of magnitude, indicating that the mechanism is not attributable to single vacancy processes.

The results of a similar analysis using the values of di-vacancies are listed in the same table. The ratios $(C_s \varepsilon M)_{exp}/M_{2V}$ are satisfactorily consistent. Due to the high concentration of vacancy defects generated by heavy deformation, vacancies primarily occur as di-vacancies rather than as isolated single vacancies.

Vacancy Complexes Formed by Deformation: A single activation process as mentioned above accounts for the variation over only two and half orders of magnitude of strain rate. However, the experimentally observed variation extends over a much wider range of strain rate of 6 orders of magnitude, extending most remarkably into the lower strain rate region. This implies the involvement of defects with less mobility or defects that dissociate with greater activation, namely small vacancy complexes larger than di-vacancies.

The experimentally observed variation in vacancy defects in gold at room temperature was reconstructed in Fig. 5 by combining five components (A, B, C, D and E) with different mobilities as shown in Fig. 6, 30, 25, 20, 15 and 10 % with associated activation energies of motion of 0.74, 0.80, 0.87, 0.97 and 1.08 eV, respectively. Of cause, this analysis is not insisting the discrete distribution of vacancy complexes, and the variation must be rather continuous as in the dotted line in the Fig. 6 from their complex variation in the size and shape.

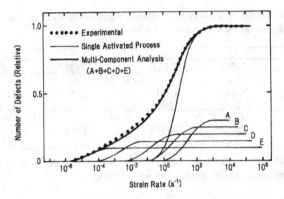

Figure 5. Multi-component fitting of the escape of vacancy defects during deformation.

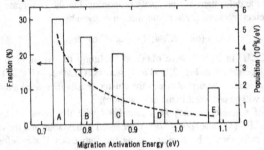

Figure 6. Activation energies of vacancy migration used in the fitting in Fig. 5.

CONCLUDING REMARKS

The formation of vacancy clusters during the deformation to fracture of thin metal foils was satisfactorily explained by the generation and reaction of dispersed vacancies. Although the vacancy cluster formation varies significantly with strain rate, vacancy defects are generated over a very wide range of deformation speeds. The results presented in this study show that the deformation without dislocation to produce large amount of vacancy defects is not the consequence of high-speed deformations, but simply depends on the specimen geometry, which enables the extreme increase of internal stress.

REFERENCES

1. M. Kiritani, Y. Satoh, Y. Kizuka, K. Arakawa, Y. Ogasawara, S. Arai and Y. Shimomura: Phil. Mag. Letters, 79 (1999) 797.
2. M. Kiritani, K. Yasunaga, Y. Matsukawa and M. Komatsu: in this volume.
3. M. Kiritani, M. Konno, T. Yoshiie and S. Kojima: Mater. Sci. Forum, 15-18 (1987) 181.
4. R. W. Balluffi: J. Nucl. Mater., 69&70 (1978) 240.

Mat. Res. Soc. Symp. Proc. Vol. 673 © 2001 Materials Research Society

Dislocation Locking by Intrinsic Point Defects in Silicon

Igor V. Peidous, Konstantin V. Loiko, Dale A. Simpson, Tony La, and William R. Frensley[1]
R&D Department, Dallas Semiconductor Corporation, Dallas, TX
[1]Department of Electrical Engineering, University of Texas at Dallas, Richardson, TX

ABSTRACT

Dislocation pileups with abnormally weak inter-dislocation repulsion have been observed in locally oxidized silicon structures. To verify if this could be attributed to elastic interaction of dislocations with intrinsic point defects, distributions of self-interstitials in dislocation stress fields have been studied using theoretical calculations and computer simulations. According to the obtained results, self-interstitials can form atmospheres about dislocations causing dislocation stress reduction and therefore screening of dislocations from interaction with external stresses. This may represent an additional mechanism of dislocation locking in silicon alternative to oxygen pinning.

INTRODUCTION

Dislocation locking in silicon was intensively studied in the 70's and was clearly demonstrated to be dependent on oxygen concentration [1, 2]. Since then, most observations of dislocation locking in silicon are traditionally attributed to their interaction with oxygen atoms and precipitates [3, 4]. Advances in silicon materials science during the past two decades have revealed an important role of Si-interstitials in defect evolution [5]. Si-interstitials are highly mobile and readily available at very high concentrations during silicon processing in device manufacturing. Being the centers of dilatation, they must interact elastically with dislocations and may cause more or less significant dislocation pinning [6]. In the present work, an estimate of dislocation locking efficiency by self-interstitials is made based on analytical calculations and computer modeling.

EXPERIMENTAL DETAILS AND CALCULATION MODELS

Dislocations were intentionally introduced in CZ (100) 150 mm silicon wafers by local oxidation. The wafers were p-type 6-9 Ohm·cm boron-doped with an oxygen concentration of 25-26 ppm (ASTM F1619). A 200 nm film of LPCVD silicon nitride was deposited and patterned using conventional lithography and plasma etch to obtain long nitride strips oriented along <110> directions on the wafer surface. Dislocation pileups were generated at the edges of the nitride film during oxidation carried out at 1100°C. After the removal of nitride and locally grown oxide films in phosphoric and hydrofluoric acids, wafers were cleaved along <110> directions perpendicular to the nitride strips. Silicon defects were delineated using preferential etching in Wright solution. Finally, dislocation distributions were studied on the wafer cross-sections using optical microscopy.

Point defect interaction with dislocations was estimated in 2-dimensional approximation of plain strains as described in detail earlier [7]. Dislocations were assumed not to act as a source or sink for point defects. Self-interstitial atoms were modeled as elastic balls. The deviation of self-interstitial concentration ΔC_0^* from its equilibrium value C_0^* without internal stresses was considered to produce specific stresses. The collective stress of self-interstitials at their equilibrium with a single dislocation having Burgers vector b_e and positioned along the z-axis in Cortesian coordinate system $\{x, y, z\}$

$$\sigma_{ij}(x, y) = \iint_S \Delta C^*(x', y') \sigma_{ij}^I(x - x', y - y') dx' dy',$$
(1)

where the integral is taken over the whole plane. Assuming that silicon volume expansion due to self-interstitials occurs only in the direction orthogonal to the dislocation, 2-dimensional effective concentration of self-interstitials is found as $C^* = 2R_iC$, where C is their volume concentration. The equilibrium interstitial concentration about a dislocation

$$C^*(x, y) = C_0^* \exp\left(-\frac{p(x, y)\Delta V}{kT}\right),$$
(2)

where k is Boltzmann's constant, T is absolute temperature, ΔV is the activation volume of a self-interstitial atom, p is the hydrostatic component of the dislocation stress, and C_0^* is the equilibrium interstitial concentration without internal stresses. A stress field of a self-interstitial in the origin of a coordinate system $\{x, y\}$ was approximated as [8]

$$\left.\begin{array}{l} \sigma_{xx}^I(x, y) = -\dfrac{2(1+v)\mu R_I (R_I - R_i)}{(1-v)[(x'+R_I)^2 + (y'+R_I)^2]}, \\[4mm] \sigma_{yy}^I(x, y) = -\sigma_{xx}^I(x, y), \\[4mm] \sigma_{xy}^I(x, y) = \dfrac{(\sigma_{xx}^I - \sigma_{yy}^I)x' y'}{[(x'+R_I)^2 + (y'+R_I)^2]}. \end{array}\right\}$$
(3)

Redistribution of point defects in dislocation stress fields was modeled using TSUPREM-4, a 2-dimensional process simulator [9]. Default values for the equilibrium concentration and diffusivity of self-interstitials were used. These coefficients are routinely employed for simulations of advanced semiconductor processes and therefore widely accepted.

RESULTS AND DISCUSSION

Preferential etching revealed extended pileup configurations of dislocations on the cross-sections of analyzed silicon wafers, as shown in Fig.1 (a). These defects formed during local oxidation of the wafers due to high stresses, which developed at the edges of nitride masking films. At the strips of nitride films orientated along <110> directions of the silicon crystal, dislocations were mainly generated in {111} planes parallel to nitride edges. They emanated from the wafer surface and penetrated into the bulk of wafers to the depth of a few hundred microns.

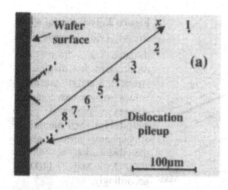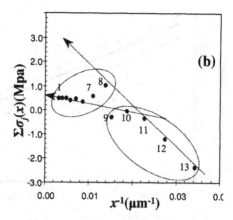

Figure 1. The optical micrograph of dislocation pileups revealed by preferential etching on a cross-section of a silicon wafer (a) and the corresponding plot of inter-dislocation repulsion versus the reciprocals of dislocation distances from the wafer surface (b).

The interaction of dislocations in the observed pileups was analyzed based on the graphical representation of the total shear stress, $\Sigma\sigma_j(x_i)$, produced by all leading dislocations in the pileup glide system, versus the reciprocals of the distances of dislocations from the effective center of their emanation ($1/x_i$) [10]. The equilibrium position x_i of i-th dislocation must satisfy the equation

$$k\frac{1}{x_i} + \sum_{j,j \neq i} \sigma_j(x_i) = \tau_c, \qquad (4)$$

where k/x_i is the stress caused by the film edge and τ_c is the critical stress of dislocation motion. However, observations showed that dislocation distributions in large pileups do not follow the linear dependency (4) of $\Sigma\sigma_j(x)$ vs. $1/x_i$. This is associated with the mechanism of pileup development. Leading dislocations form and move at higher temperatures of an oxidation process and thus experience lower τ_c. [11].

A characteristic feature observed in the dislocation distribution was that last dislocations in a leading series and first dislocations in a following up series, like 8th and 9th dislocations in Fig.1 (a), demonstrated an abnormal interaction. The analysis of inter-dislocation repulsion (Fig.1, (b)) suggests that during the formation of follow up dislocations, leading dislocations are already pinned and their stresses are weakened. Dislocations exhibited a weak interaction regardless of the wafer type, thermal history and oxygen concentration. Such a week repulsion of dislocations can be attributed to their interaction with fast-diffusing defects, such as intrinsic point defects in silicon.

To validate this assumption, a potential influence of point defects on dislocation stress fields was estimated theoretically. The authors considered interaction of self-interstitial atoms in silicon with a dislocation [7]. According to numerical calculations performed, a dislocation causes redistribution of self-interstitials in a way that dislocation stresses are significantly reduced. Fig. 2 shows the shear stress of an edge dislocation as a function of the distance x from it. The dislocation is oriented along z-direction and the distance x is counted out in the direction of its Burgers vector.

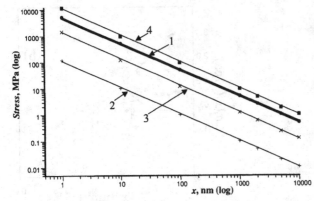

Figure 2. The level of shear stress $\tau_{xy}(x,0)$ versus the distance x from the dislocation: 1 - dislocation stress; 2, 3 and 4 – absolute values of negative shear stresses produced by equilibrium self-interstitial atmospheres formed about the dislocation at 900°C, 1000°C, and 1100°C, accordingly.

The stresses produced by a Cottrell atmosphere are reverse to the dislocation stress. Thus, collective stresses of point defects compensate the dislocation stress. Such a reduction in stresses leads to screening of the dislocation from interaction with other defects. Along with the increase in the background equilibrium concentration of self-interstitials in silicon with temperature, the stress produced by a Cottrell atmosphere becomes greater at higher temperatures.

The calculation was not self-consistent with respect to stress reduction around a dislocation during self-interstitial redistribution. Also, inter-interstitial repulsion was not taken into account. Hence, the effect of dislocation stress compensation by self-interstitials is overestimated. On the other hand, vacancies and impurities can participate in the formation of a point defect atmosphere around a dislocation and contribute to its stress reduction. Additionally, many processes of semiconductor device fabrication, such as diffusion and oxidation, provide point defects in excess of their equilibrium concentration [12,13] that should enhance dislocation screening. In any case, the results of calculations support the assumption that intrinsic point defects can screen dislocations. In addition to Peierls stress for "clean" dislocations and the stress of breakaway from pinning points and/or atmospheres, the unlocking stress for a screened dislocation includes the increment required to compensate for the weakening of dislocation stresses.

Due to their exceptional mobility in silicon, self-interstitials may cause the most rapid dislocation screening. Computer simulations showed that at 1100°C, the saturation time for the concentration of self-interstitials at distances R_{atm} of a few hundred angstroms from a dislocation is only a fraction of a second (Fig.3). At the same time, it may be as long as 0.5 hour for $R_{atm} \approx 1$ μm. It follows from Fig.3 that a dislocation starts to lose its Cottrell atmosphere of self-interstitials at $R_{atm} \approx 250$ Å only if it moves with a speed greater than a few μm/s.

The simulations also showed that even for dislocations moving with the speed $V \leq 10$ μm/s, self-interstitials establish notable quasi-equilibrium Cottrell atmospheres moving along with dislocations (Fig.4). Thus, due to their extreme mobility in silicon, self-interstitial atmospheres may not exhibit significant dragging forces for dislocations gliding with a speed lower than 1 μm/s at high temperatures.

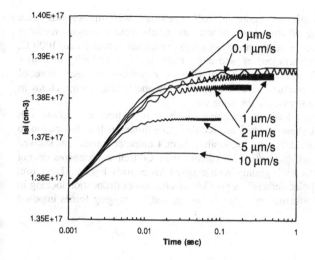

Figure 3. The simulation results on the concentration of self-interstitials at 250 Å above a dislocation versus the dislocation lifetime (time after formation) for varying speed of dislocation movement.

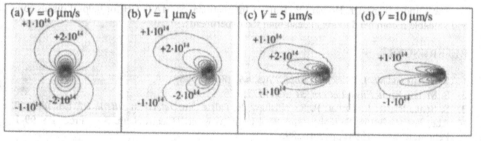

Figure 4. The contour plots representing simulation results on quasi-equilibrium concentrations of self-interstitials about dislocations gliding with different velocities at 1100°C. The concentrations are given in terms of their deviation from the background concentration of self-interstitials. The contours are plotted with the step of 10^{14} at/cm.$^{-3}$

Dislocations that may form in silicon wafers during device manufacturing are reported to move with $V \leq 1$ μm/s [14]. Therefore, according to the simulation results, their movement is always subjected to the influence of self-interstitials by means of screening.

CONCLUSIONS

Distributions of dislocations in pileup configurations generated during local oxidation of silicon wafers were studied. Theoretical calculations of the equilibrium dislocation distributions and critical stresses of dislocation movement suggested that dislocations formed at earlier stages of the oxidation process exhibited weaker interaction with other defects. The authors assumed that this was associated with screening of dislocation stress fields by fast-diffusing point defects, in particular, by self-interstitials.

Numerical calculations of elastic interaction of self-interstitials with dislocations were performed. They showed that a Cottrell atmosphere of self-interstitials about a dislocation causes dislocation stress compensation. At the temperatures of silicon processing greater than 1000°C, the background equilibrium concentration of self-interstitials is high enough to reduce dislocation stresses drastically. This result agrees well with the experimental observations of dislocation distributions in locally oxidized silicon structures. Vacancies and foreign atoms in silicon should influence dislocation stresses in the same way.

Reduction of dislocation stresses leads to screening of dislocations from interaction with external stresses. Higher resolved shear stresses are required to enforce such a dislocation to move. Thus, the screening of dislocations does immobilize them. Computer simulations showed that due to extreme mobility of self-interstitials in silicon, their Cottrell atmospheres do not create significant barriers for dislocation gliding with a speed lower than 1 μm/s. Therefore, dislocation screening by intrinsic point defects is a possible mechanism of dislocation locking in addition to that associated with the formation of pinning points and/or dragging forces imposed by Cottrell atmospheres.

ACKNOWLEDGMENTS

The authors thank Dr. Howard R. Huff of International SEMATECH for helpful discussions and valuable recommendations in designing the experiments.

REFERENCES

1. S. M. Hu and W. J. Patrick, *J. Appl. Phys.* **46**, 1869 (1975).
2. S. M. Hu, *Appl. Phys. Letters,* **31**, 53 (1977).
3. S. Senkader, K. Jurkschat, P. R. Wilshaw, R. Falster, in *Defects in Silicon III*, Edited by T. Abe, W. M. Bullis, S. Kobayashi, W. Lin, P. Wagner, (Electrochem. Soc. Proc. **PV 99-1**, Pennington, NJ, 1999) pp. 280-289.
4. M. Akatsuka, K. Sueoka, H. Katahama, N. Morimoto and N. Adachi, *Jpn. J. Appl. Phys., Part 2*, **36**, L1422 (1998)
5. T. E. Haynes, *MRS Bulletin*, **25**, 14 (2000).
6. J. P. Hirth, J. Lothe, *Theory of Dislocations*, (McGraw-Hill, New York, 1968) pp. 456, 633.
7. I. V. Peidous and K. V. Loiko, in *High Purity Silicon VI*, Edited by C. L. Claeys, P. Rai-Choudhury, M. Watanabe, P. Stallhofer, H. J. Dawson, (Electrochem. Soc. Proc. **PV 2000-17**, Pennington, NJ, 2000) pp. 145-155.
8. L. D. Landau, E. M. Lifshitz, *Theory of Elasticity*, (Butterworth-Heinemann, Oxford, 1997) p. 19.
9. *TSUPREM-4 User's Manual, Version 2000.4*, Avant! Corporation, (Fremont, CA, December 2000).
10. S. M. Hu, *Appl. Phys. Lett.*, **31**, 139 (1977).
11. I. V. Peidous, C.H.Gan, R. Sundaresan, S. K. Lahiri, in *High Purity Silicon V*, Edited by C. L. Claeys, P. Rai-Choudhury, M. Watanabe, P. Stallhofer, H. J. Dawson, (Electrochem. Soc. Proc. **PV 98-13**, Pennington, NJ, 1998) pp. 272-281.
12. H. Park, K. S. Jones, J. A. Slinkman, M. E. Law, *J. Appl. Phys.*, **78**, 3664 (1995).
13. R. Y. S. Huang and R. W. Dutton, *J. Appl. Phys.*, **74**, 5821 (1993).
14. H.Shimizu, S.Isomae, K.Minowa, T.Sotoh, T.Suzuki, *J. Electrochem. Soc.*, 145, 2523 (1998).

Mat. Res. Soc. Symp. Proc. Vol. 673 © 2001 Materials Research Society

Optical study of SiGe films grown with low temperature Si buffer

Y. H. Luo, J. Wan, J. L. Liu, and K. L. Wang
Device Research Laboratory, Department of Electrical Engineering,
University of California at Los Angeles, Los Angeles, CA 90095-1594

ABSTRACT

In this work, SiGe films on low temperature Si buffer layers were grown by solid-source molecular beam epitaxy and characterized by atomic force microscope, photoluminescence and Raman spectroscopy. Effects of the growth temperature and the thickness of the low temperature Si buffer were studied. It was demonstrated that using proper growth conditions of the low temperature Si buffer, the Si buffer became tensily strained and gave rise to the compliant effect. High-quality SiGe films with low threading dislocation density have been obtained.

INTRODUCTION

High-quality strain-relaxed SiGe buffer layers have been widely used as "virtual substrates" for the growth of strain Si/SiGe high electron mobility transistors and metal-oxide-semiconductor field effect transistors [1,2], and Ge photodiodes on Si [3]. However, the large lattice mismatch (~ 4.17%) between Si and Ge usually results in lots of threading dislocations in SiGe buffer layers, which propagate through the SiGe buffer layer into the active layers and deteriorate the device performance [4]. Several methods have been used to grow high quality strain-relaxed SiGe, such as graded composition [5], compliant substrate [6], etc. Recent reports indicated that the use of low-temperature (LT) buffer layers could significantly reduce threading dislocation density in the SiGe layer [7]. It was believed that the LT Si layer plays important roles: provides low energy sites for dislocation nucleation, or point defects for trapping of propagating dislocations, and contain the mismatch strain [8]. In this work, photoluminescence (PL) and Raman spectroscopy were used to study the strain and dislocations of the relaxed SiGe films grown on low temperature Si buffer.

EXPERIMENTAL DETAILS

The samples investigated were grown by solid source molecular beam epitaxy (MBE) in a Perkin-Elmer system. A P$^-$ type Si (001) substrate was cleaned by using a modified Shiraki method and loaded into the MBE system after a diluted HF dip. At first, a 60 nm Si buffer was grown at 600 $^\circ$ C to reduce the defects and smooth the surface of the substrate. After the growth of a followed-up LT Si layer, a SiGe layer was grown at 480°C. The growth rate for both the Si layer and the SiGe layer were about 1Å/s. PL was measured at 4.5 K using an Ar$^+$ 488 nm laser line. Raman spectra were taken using a 457.9 nm Ar$^+$ laser line.

Two series of samples were studied in this paper. Series A is 150 nm Si$_{0.8}$Ge$_{0.2}$ films grown on 200 nm LT Si buffer layer with different growth temperatures, which was used to optimize the growth temperature of the LT Si buffer layer. Series B is 200 nm Si$_{0.8}$Ge$_{0.2}$ films grown on 400°C LT Si layer with different thickness, which was used to optimize the thickness of the LT Si buffer layer.

RESULTS AND DISCUSSION

PL spectra from the series A samples with different growth temperatures are shown in Fig. 1. The peaks were normalized referred to the Si TO peak from the Si substrate. Two main peaks are observed around 0.8eV and 0.9eV, the D1 and D2 line (D1, D2), which were believed to be from the dislocation intersections [9,10]. In the inset, the integrated intensity of the D1/D2 peaks versus the growth temperature are shown. The largest intensity was obtained for 400°C. The reason for this phenomena could be following: point defects in the LT Si buffer layer worked as nucleation centers to form dislocations and defects in the LT Si buffer layer, instead of in the SiGe layer. The defects degraded the PL intensity of the Si transverse optical (TO) peak from the Si substrate. As a result, the higher D1/D2 peak intensity implied the higher defect density in the LT Si buffer layer. So for the 400°C growth temperature, the dislocation and defect density in the LT Si buffer layer was the highest and as a result, lest dislocations formed in the SiGe layer. The smallest surface roughness and ordered cross-hatch pattern from the atomic force microscope measurement confirmed this assumption.

The Raman spectra of series A are shown in Fig. 2(a). By using an iterated Lorentzian fit to the curves, the peak position and full width at half maximum (FWHM) were obtained. The position and the FWHM of the Si-Si peak from the SiGe layer with different temperature buffers are shown in Fig. 2(b). It is known that the stress only shifts the peak without changing the FWHM of the peak while threading dislocations and defects make the Si-Si peak redshift and wider [10,11].

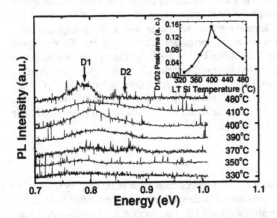

Figure 1. Low energy parts of photoluminescence spectra the 150 nm $Si_{0.8}Ge_{0.2}$ samples on the low temperature Si buffer grown at different temperatures (series A). The peaks labeled as D1 and D2 are dislocation lines. The inset shows the integrated intensity of D1 and D2 peak normalized to the Si transverse optical peak from the Si substrate.

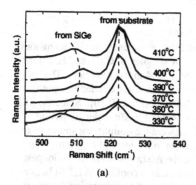
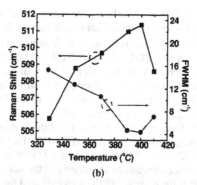

Figure 2. (a) Raman spectra of the 150 nm Si$_{0.8}$Ge$_{0.2}$ samples on the low temperature Si buffer grown at different temperatures (series A). Two peaks correspond to the Si-Si peak from the SiGe layer (~ 510 cm^{-1}), and from the Si buffer and substrate (~ 522 cm^{-1}), respectively. (b) Peak position and full-width at half maximum of the Si-Si peak from the SiGe layers versus the growth temperature of the low temperature Si buffer.

The peak position and FWHM of the Raman results implied that the defect density of the SiGe film in the sample with a 400°C LT Si buffer was the lowest. The sample with 410°C LT Si buffer was just a little rougher as observed by AFM. However, the Raman peak redshifted substantially and the FWHM was much larger, indicating that there was a high density of defects in the SiGe film. For the sample with a 390°C LT Si buffer, though no ordered crosshatches were observed, since the Raman shift and FWHM was very close to that of the sample with a 400°C buffer, the crystalline quality of this SiGe film was believed to be quite closed to that of the 400°C sample. For samples with a LT Si buffer grown at a temperature around 400°C, the quality of the SiGe layer was the best, while for too low and too high temperatures, the quality deteriorated.

The PL spectra from 200nm Si$_{0.8}$Ge$_{0.2}$ on different thick LT Si buffers (series B) are shown in Fig. 3. In Fig. 3(a), the high-energy parts of PL spectra from the samples are shown. For samples A and B, a shoulder to the left of the strong Si TO peak from the substrate was observed (as indicated by dash lines). For samples C, D, E and F, the peak became stronger and separated from the major Si TO peak. Using Gaussian fits to these two peaks, the positions of the peaks were determined. The TO Si peak from the Si substrate remains at 1.092 eV, while the left peak changed position with the thickness of the LT Si buffer layer. The inset in Fig. 3(a) shows the shift of the left peak with respect to the Si TO peak as a function of the thickness of the LT Si buffer. At first (A-C), the redshift of the peak increases with the thickness of LT Si buffer layer, reaching a maximum value for the 200 nm LT Si buffer sample (D), then decreases slowly (E-F). It is believed that this peak is the Si TO peak from the LT Si buffer and the tensile strain of the LT Si buffer causes the redshift [13]. The maximum redshift for sample D indicates the highest tensile strain of the LT Si buffer. From the relationship between the energy bandgap change and strain of the strain Si on Si$_{1-x}$Ge$_x$ $E_g(Si)$-0.4 [13], the strain may be estimated from the redshift. For sample D, the shift of 20 meV corresponds to the strain Si on Si$_{0.95}$Ge$_{0.05}$. Since the Ge composition of the SiGe layer on LT Si layer is about 0.18, the strain of the LT Si layer is about 28% of the fully mismatched strain between the Si and the $Si_{0.82}Ge_{0.18}$. Partially because of this accommodation, the misfit dislocation density decreased.

The low energy parts of PL spectra are shown in Fig. 3(b). For clarity, the PL intensities are not to scale. A peak (D4) at about 0.9 eV was observed for samples B, C, D, and E, but did not show up in samples A and F. This peak was attributed to the D4 dislocation line. The threading dislocations in the SiGe layers, the Si buffer and substrate, and extended straight segments of misfit dislocations are responsible for the D4 dislocation line [9,10]. Since the SiGe layer in our samples was only 200 nm thick, the straight segments of misfit dislocations and threading dislocations in the LT Si buffer and the Si substrate were the dominating source of the D4 line. The integrated intensity ratio of D4/(D1+D2) is shown in the inset of Fig. 3(b). The highest D4/(D1+D2) intensity ratio in sample D implies that the misfit dislocations are longest with the lest intersections and/or the density of threading dislocations confined in LT Si buffer and Si substrate is the highest. Both cases indicated the lowest threading dislocation density in the SiGe layer. Our AFM measurement results indicated long straight misfit dislocations in sample D [8].

The samples of series B were measured by Raman spectroscopy for the strain relaxation and film quality. The peak position and the FWHM of the Si-Si peak of the SiGe layer with different thicknesses of buffers are shown in Fig. 4. The Raman results implied that the defect density of the SiGe film with a 200 nm LT Si buffer was the lowest.

For the sample with a 250 nm LT Si buffer, even though no ordered crosshatch was observed [8], as indicated the Raman shift and FWHM, the threading dislocation density and crystal quality of this SiGe film were quite close to the 200 nm LT Si sample. For the sample

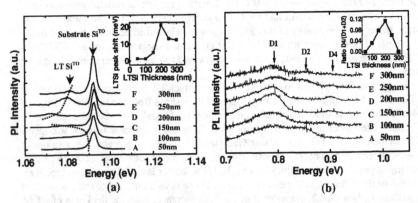

Figure 3. (a) High energy parts of photoluminescence spectra from series B samples. The peak to the left of the Si transverse optical peak of the substrate came from the low temperature Si buffer layer. The inset shows the shift value of the left peak (transverse optical of low temperature Si) with respect to the substrate Si transverse optical peak. (b) Low energy parts of photoluminescence spectra from sample A through F. The peaks are labeled as D1, D2, and D4 dislocation lines, respectively. The inset shows the integrated intensity ratio of the D4 peak to the D1 and D2 peak D4/(D1+D2).

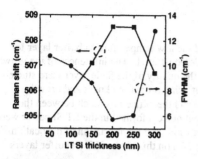

Figure 4. Peak position and full-width at half maximum of the Si-Si peak from the SiGe layers versus the thickness of the low temperature Si buffer for 200 nm $Si_{0.8}Ge_{0.2}$ on a low temperature Si buffer grown at 400°C (series B).

With 300nm LT Si buffer was just little rougher [8]. However, the Raman peak redshifted substantially and the FWHM was much larger, which indicated a high density of defects. For the samples with the thickness of LT Si buffer below 100 nm, the Raman results indicated much poorer quality, which matched AFM and DAXRD results [8].

In order to study the nature of the LT Si buffer layer, the Raman spectra for a 50 nm and a 200 nm 400°C LT Si buffer samples without a SiGe layer are shown in Fig. 5. For comparison, the spectrum for a 200 nm Si buffer layer grown at 500°C is also shown. For the 50 nm LT Si buffer, there was a tail to the left side of the Si-Si peak of the LT Si buffers coming from the point defects in the LT Si layer. However, for the 200 nm LT Si buffer, in addition to a larger tail to the left, the Si-Si peak redshifted about 1.5 cm^{-1} and the FWHM was much larger than that of the 50 nm LT Si buffer. The later two facts were attributed to the disorder and point defects in the LT Si buffer. This indicated that the 200 nm LT Si buffer had a much higher density of point defects than the 50 nm LT Si buffer, and these point defects worked to improve the quality of the SiGe layer.

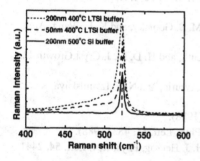

Figure 5. Raman spectra from low temperature Si buffer layers without SiGe layer. From top to bottom: 200 nm 400°C low temperature Si buffer layer, 50 nm 400°C low temperature Si buffer layer, and 200 nm 500°C low temperature Si buffer layer. The dash straight line indicates the peak position of the unstrained Si-Si line.

CONCLUSIONS

In conclusion, the compliant substrate effect of the low temperature Si buffer layer for growth of high quality thin relaxed SiGe layers was demonstrated. The influences of the growth temperature and the thickness of the LT Si buffer on the quality of the SiGe layer were studied using PL and Raman spectroscopy. It was shown that using proper thickness and growth temperature, the LT Si buffer became tensily strained and the lattice mismatch between the buffer layer and the SiGe layer was reduced. The point defects formed in the LT Si buffer were believed to enhance the nucleation of the misfit dislocations and the formation of dislocations in the LT Si layer. Thus the quality of the SiGe films grown on this kind of LT Si buffer layers were improved.

ACKNOWLEDGEMENTS
The work was supported by Semiconductor Research Corporation and UCMICRO-Conexant.

REFERENCES

1. K. Ismail, F. K. LeGoues, K. L. Saenger,M. Arafa, J. O. Chu, P. M. Mooney, and B. S. Meyerson, Phys. Rev. Lett. **73**, 3447 (1994).
2. Y. H. Xie, D. Monroe, E. A. Fitzgerald, P.J. Silverman, F. A. Theil, and G. P.Watson, Appl. Phys. Lett. **63**, 2263 (1993).
3. L. Colace, G. Masini, G. Assanto, H.-C. Luan, K. Wada, and L.C. Kimerling, Appl. Phys. Lett. **76**, 1231 (2000).
4. R. Hull, J. C. Bean, and C. Buescher, J. Appl. Phys. **66**, 5837 (1989).
5. J. L. Liu, C. D. Moore, G. D. U'Ren, Y. H. Luo, Y. Lu, G. Jin, S. G. Thomas, M. S. Goorsky, and K. L. Wang, Appl. Phys. Lett. **75**, 1586 (1999).
6. Y. H. Luo, J. L. Liu, G. Jin, K. L. Wang, C. D. Moore, M. A. Goorsky, C. Chih, and K. N. Tu, J. Electron. Mater. **29**, 950 (2000).
7. H. Chen, L. W. Guo, Q. Cui, Q. Hu, Q. Huang, and J. M. Zhou, J. Appl. Phys. **79**, 1167, (1996).
8. Y. H. Luo, J. Wan, R. L. Forrest, J. L. Liu, G. Jin, M. S. Goorsky, and K. L. Wang, Appl. Phys. Lett. **78**, 454 (2001).
9. H. P. Tang, L. Vescan, C. Dieker, K. Schmidt, H. Luth, and H. D. Li, J. Cryst.Growth **125**, 301 (1992).
10. E. A. Steinman, V. I. Vdovin, T. G. Yuhova, V. S. Avrutin, and N. F. Izyumskaya, Semicond. Sci. Technol. **14**, 582 (1999).
11. J. Takahashi, and T. Makino, J. Appl. Phy. **63**, 87 (1988).
12. D. J. Olego, H. Baumgart, and C. K. Celler, Appl. Phys. Lett. **51**, 483 (1988).
13. G. Abstreiter, H. Brugger, T. Wolf, H. Jorke, and H. J. Herzog, Phys. Rev. Lett. **54**, 2441 (1985).

Mat. Res. Soc. Symp. Proc. Vol. 673 © 2001 Materials Research Society

Thickness-fringe Contrast Analysis of Defects in GaN

Jeffrey K. Farrer, C.Barry Carter, Z. Mao, Stuart McKernan
Dept. of Chemical Engineering and Materials Science, University of Minnesota
421 Washington Ave. S.E., Minneapolis MN 55455 USA

ABSTRACT

The analysis of thickness-fringe contrast in weak-beam transmission electron microscope (TEM) images has been shown to be a reliable method for the complete determination of the character, as well as the magnitude, of a dislocation Burgers vector. By selecting multiple diffraction conditions and, for each condition, determining the number of terminating thickness fringes at the exit of a dislocation from a wedge-shaped sample, the Burgers vector can be unambiguously determined. Defect analysis of GaN pyramids grown on (111)Si by the lateral epitactic overgrowth (LEO) technique reveals a core region which contains a relatively high density of dislocations and a lateral-growth region where the defect density is decreased. The thickness-fringe contrast technique was used in the lateral growth regions of the pyramids to analyze the dislocation Burgers vectors.

INTRODUCTION

The performance of microelectronic and optoelectronic devices is often directly related to the type (i.e. Burgers vector) and density of dislocations in the III–V materials used in the device. There are several methods of analyzing dislocation Burgers vectors [1]. One of the most popular of these techniques is based on the "invisibility criterion" ($g \bullet b = 0$), which under strong two-beam conditions can determine the direction of the Burgers vector. However, if the dislocation contains a large edge component, the $g \bullet b = 0$ condition does not accurately describe the dislocation contrast since there may be a displacement of the lattice that is perpendicular to the Burgers vector (i.e. glide-plane buckling). There are also complications when using a high-order diffracting vector and when the structure of the material is complex.

A method has been developed which overcomes these disadvantages and which allows for the determination of both the sign and the magnitude of the Burgers vector. [2]. This method uses weak-beam conditions in the TEM to image the thickness fringes near a dislocation. The complete determination of the magnitude and direction of the dislocation is accomplished by counting the number of extra equal-thickness fringes, n, observed around a dislocation outcrop. The number of fringes is then applied to the relation

$$n = g \bullet b$$

where g represents the diffracting vector. By selecting three diffracting vectors, g, and determining n for each different condition, b can be unambiguously determined.

EXPERIMENTAL DETAILS

As part of this study the technique described above was used to analyze the Burgers vectors of dislocations found in GaN structures grown on GaN/AlN/Si(111) substrates. The GaN structures were grown by the LEO technique [3]. The lateral overgrowth was controlled by depositing a Si₃N₄ mask on the GaN layer. Apertures 5 μm in diameter and 20 μm apart were etched into the mask to act as seeding locations for further GaN growth. This additional growth resulted in the formation of GaN pyramids above the apertures in the patterned Si₃N₄ mask. The growth was carried out in a low-pressure metal organic chemical vapor depostion (MOCVD) system at 1050°C for approximately 3 hours. Further details on the growth of the GaN are given elsewhere [4].

TEM specimens were prepared by gluing two pieces of the sample together with epoxy and cutting and polishing the whole structure in cross section. The thin cross sections are further polished and dimpled to a thickness of about 10 μm. The specimens are then thinned to electron transparency by ion milling. Observation of the final TEM foils was carried out in a Philips CM30 TEM operating at 300 kV. The morphology of the as-grown structures was studied by secondary electron imaging in a SEM.

RESULTS AND DISCUSSION

A secondary-electron image of the as-grown structures, shown in Figure 1a, indicates that each pyramid has predominantly six $\{10\bar{1}1\}$ facets. The width of the base and the height of the pyramids are both approximately 15 μm. Schematics of the pyramid structure are shown in Figures 1b and 1c. Figure 1b represents a top view of the pyramid which shows the location of the aperture in the Si₃N₄ mask, above which the core region of the pyramid is grown. It also shows the lateral growth portion of the pyramid surrounding the core. The vertical bar indicates from approximately where in the pyramid the TEM cross-section sample (shown in Figure 3) was taken. On the right is a schematic representing a cross-section view of the pyramid which shows the GaN/AlN seeding layer and the (111)Si substrate. The core region is shown containing vertical lines which represent the threading dislocations that start at the seed layer [5]. The lateral growth region, outside the core, contains horizontal lines which represent dislocations that run parallel to the growth interface.

Selected-area diffraction (SAD) was performed in the TEM on the interface area, the GaN pyramid and the substrate. The SAD patterns confirmed that the GaN pyramids, the GaN seed layer, and the AlN buffer layer are all monocrystalline with the following epitactic relationship with respect to the Si substrate :

$$[11\bar{2}0]_{GaN} \| [11\bar{2}0]_{AlN} \| [\bar{1}10]_{Si}$$
$$(0001)_{GaN} \| (0001)_{AlN} \| (111)_{Si}$$

Figure 2 is a low magnification weak-beam dark-field (WBDF) image of the cross-section specimen. The image was obtained with the electron beam nearly parallel to the [11$\bar{2}$0] GaN zone axis using a g/3g condition, g = [1$\bar{1}$00] . In the core region of the pyramid, dislocations thread nearly perpendicular to the interface plane. The density of dislocations in this region appears to be higher than any other region of the pyramid. To the sides of the core (the lateral growth region), the dislocations can be seen to lie parallel to the interface.

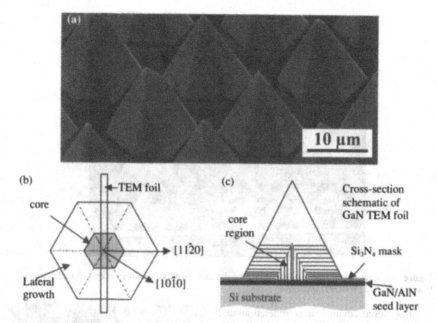

Figure 1. A secondary-electron SEM image of the as-grown GaN pyramids (a). Schematics representing a top view (b) and a cross section (c) of the pyramid structure.

Figure 3 is a WBDF image at a higher magnification than Figure 2. The image was obtained with the electron beam nearly parallel to the [11$\bar{2}$0] GaN zone axis, but using a g/5g condition, g = [0002]. This image reveals the microstructure near the boundary of the core region of the pyramid. The dislocations can clearly be seen threading up through the core region and then turning abruptly, at the inclined boundary of the core, to lie parallel to the substrate. The location of the "bending" of the dislocations indicates approximately where the lateral growth of the pyramid begins.

The thickness-fringe contrast technique for Burgers vector analysis is based on a two-beam diffraction condition in the TEM [6]. Under these conditions, the Howie–Whelan equations may be used to describe the amplitudes of the transmitted and diffracted beams [1]. Using the "modified" Howie–Whelan equations (modified to account for the presence of a strain field), the following equation can be derived, which describes the behavior of thickness fringes near a dislocation. [7]

$$ts_{eff} = ts + \mathbf{g} \cdot [\mathbf{R}(t) - \mathbf{R}(0)] = constant \quad (1)$$

where t is the local thickness of the sample, s is the deviation from the exact Bragg condition, s_{eff} is the deviation from the Bragg condition near a displacement field, and $\mathbf{R}$ is the displacement field near a dislocation.

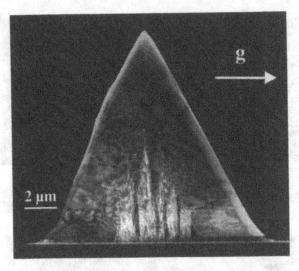

Figure 2. A low magnification TEM image of the cross-section specimen acquired under weak-beam, dark-field (WBDF) conditions.

If a closed circuit L is considered around the dislocation line, the displacement vector $\mathbf{R}$, taken along the circuit L in a left-hand sense, is $\mathbf{R}|_L = -\mathbf{b}$. The first term on the right-hand side of equation (1), ts, and the third term, $-\mathbf{g} \cdot \mathbf{R}(0)$, are zero after integration about the closed circuit. However, the second term $-\mathbf{g} \cdot \mathbf{R}|_L = -\mathbf{g} \cdot \mathbf{b}$ is only zero if $\mathbf{g}$ and $\mathbf{b}$ are perpendicular. The left-hand side of equation (1) taken around the circuit L is $ts_{\text{eff}|L} = n$, which is the difference in the number of fringes that exit and the number that enter the closed circuit [6]. Therefore, under weak-beam conditions, the number of terminating fringes n at the intersection of the dislocation with the free surface is equal to $\mathbf{g} \cdot \mathbf{b}$.

Figure 4 is shown to illustrate the method of using the thickness-fringe contrast to determine the Burgers vectors of dislocations. All of the WBDF images in Figure 4 are of the same region, but each was acquired using different diffraction conditions. The diffracting vector $\mathbf{g}$ is recorded below each image and the direction of $\mathbf{g}$ is indicated in the image. There are four dislocations (or dislocation loops) labeled A, B, C and

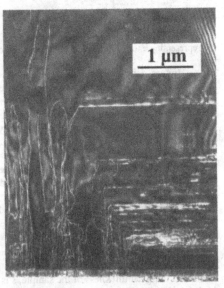

Figure 3. A WBDF image of the boundary of the core region of the pyramid.

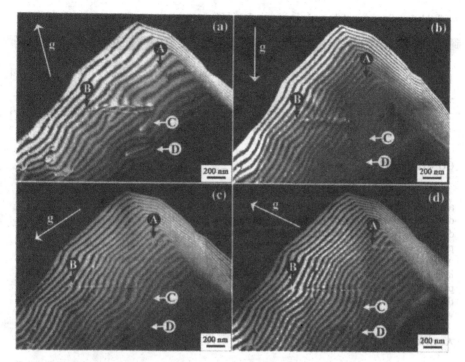

Figure 4. WBDF images of four different dislocations or dislocation loops labeled A, B, C, and D in each image. The diffraction conditions are **g**/5**g**, with **g** equal to (a) $0\bar{1}13$, (b) 0002, (c) $1\bar{1}0\bar{1}$, and (d) $1\bar{1}01$.

D. The thickness fringes terminating at the ends of the dislocations can be used to determine the Burgers vectors of the dislocations. Both ends of all the dislocations can be seen clearly, however the thickness fringes which terminate on the left are used in the analysis. If the terminating fringe runs toward the top of the image the number of fringes, n, is positive, otherwise n is negative.

The results of the analysis are summarized in Table 1. A total of four different diffraction conditions are presented here although only three are needed for the Burgers vector analysis. The line directions of the dislocations, which are also reported in the table, were obtained by trace analysis. From these results it has been determined that defect A is a screw dislocation, while defect B appears to be a dislocation half loop lying on the basal plane. Defects C and D are mixed dislocations; the character of these defects being that of 30° and 60° dislocations, respectively. One advantage of this technique is that analysis of the Burgers vector is not dependent upon the contrast of the dislocation or on the need to recognize the condition **g** • **b** = 0. It can also be recognized, from the table, that a maximum of three non-coplanar diffraction conditions are needed for the Burgers vector analysis to be complete.

Table 1. A Summary of the analysis from thickness fringes terminating at dislocations in Figure 4

g	zone axis	A	B	C	D
g = 0 1̄13	[112̄1]	0	0	-1	0
g = 0002̄	[011̄0]	0	0	0	0
g = 1 1̄0 1̄	[112̄0]	-1	-1	0	-1
g = 1̄ 1̄01	[112̄0]	-1	-1	0	-1
Burgers vector		1/3[2̄110]	1/3[12̄10]	1/3[1̄1̄20]	1/3[12̄10]
Line direction		[2̄1̄10]	loop	[1̄010]	[2̄1̄10]

CONCLUSION

The thickness-fringe contrast technique for Burgers vector analysis has been shown to be a reliable method for the complete determination of the character, as well as the magnitude, of a dislocation Burgers vector in GaN. By selecting at least three diffracting conditions, **g**, and determining n for each condition, **b** has been unambiguously determined. Defect analysis of GaN pyramids grown on (111)Si by the LEO technique reveals a core region within the pyramids, which contains a relatively high density of threading dislocations. As these dislocations reach the lateral growth region of the pyramid they bend in such a manner that their line direction becomes parallel to the substrate/epilayer interface. The thickness-fringe contrast technique was used in the lateral growth regions of the pyramids to observe the presence of dislocations with Burgers vectors equal to $^1/_3<112̄0>$.

ACKNOWLEDGEMENTS

The authors gratefully acknowledge W. Yang and S.A. McPherson at the Honeywell Technology Center (HTC) for the GaN growth. This research is supported by the NSF under contracts NSF/DMR 9522253 and NSF/DMR 0102327 with partial support from HTC.

REFERENCES

1. D.B. Williams, Carter, C. B., *Transmission Electron Microscopy-A Textbook for Materials Science* (Plenum Press, New York, 1996).
2. Y. Ishida, H. Ishida and K. Kohra, Fifth International Conference on High Voltage Electron Microscopy, 623-6 (1977).
3. O. Nam, M.D. Bremser, B.L. Ward, R.J. Nemanich and R.F. Davis, Japanese Journal of Applied Physics **36** (5A), L532 (1997).
4. Z. Mao, McKernan, Carter, C. B., Yang, W., McPherson, S. A., MRS Internet J. of Nitride Semicond. Res. 4S1, G3.13 (1999).
5. Z. Mao, M.T. Johnson and C.B. Carter, Microscopy and Microanalysis **4** (Suppl. 2), 628-9 (1998).
6. Y. Ishida, Ishida, H., Kohra, K., Ichinose, H., *Phil. Mag. A* **42**, 453-62 (1980).
7. F.W. Schapink, Phys. Stat. Sol. A **29**, 623-34 (1975).

Dislocations in
Small Structures

Mat. Res. Soc. Symp. Proc. Vol. 673 © 2001 Materials Research Society

Modeling of Dislocations in an Epitaxial Island Structure

X. H. Liu, F. M. Ross and K. W. Schwarz
IBM Watson Research Center, P.O. Box 218
Yorktown Heights, NY 10598, U.S.A.

ABSTRACT

We present calculations of dislocations in $CoSi_2$ islands grown by reactive epitaxy on a Si(111) substrate. The stress fields due to the lattice mismatch are calculated with standard FEM techniques, and are converted into a structured, multi-level and multi-grid stress table that is imported into the PARANOID code to study the dislocation dynamics. Single and multiple dislocations in the island have been simulated, and the predicted patterns are strikingly similar to those observed experimentally. By looking at the growth behavior of very small loops we also find that dislocation-loop nucleation becomes easier as the islands become larger, and that thick islands are dislocated at smaller sizes than thin ones. These results are also in good agreement with experimental observations. We conclude that current modeling techniques are sufficient to treat this type of problem at a useful level of accuracy.

INTRODUCTION

Dislocations are linear defects in crystals which mark the places where atomic planes have slipped relative to each other. The nucleation and motion of these defects influence such mechanical properties as strength, hardness, and fracture toughness. Because of this, dislocations have historically been studied with an emphasis on the mechanical properties of bulk materials. More recently, however, dislocations have become a topic of interest in the manufacture of semiconductor films and devices, where they provide electrical leakage paths which can ruin device performance. In such a situation, the task is to understand the behavior of just a few dislocations in a highly confined geometry. As a paradigm of such a problem, we examine the growth of dislocations in certain island structures which form spontaneously during epitaxial growth. Such dislocated islands provide an ideal environment in which to study dislocations in small structures. A preliminary report on this work has been published elsewhere [1].

In the semiconductor industry, device structures are fabricated through the repeated application of a number of basic processing steps involving epitaxy, photolithography, deposition, etching, oxidation, diffusion, ion implantation, evaporation or sputtering, and chemical-mechanical polishing. Many of these processes generate large stresses, which may initiate and drive dislocations into the device. In contrast to bulk materials, the stresses in devices are often caused, not by mechanical loads, but by mismatches in physical properties such as thermal expansion and the varying lattice constants of the various materials used in device fabrication. Although the dislocation density is comparable for bulk materials and semiconductor structures, the number of dislocations in devices is significantly fewer because of their small size. This renders continuum plasticity theory, developed to study the plastic behaviors of bulk materials with millions of dislocations, inapplicable to semiconductor structures where only a few dislocations appear, often in a very complicated environment. For the latter, one must resort to the discrete modeling of dislocations if one wishes to make progress.

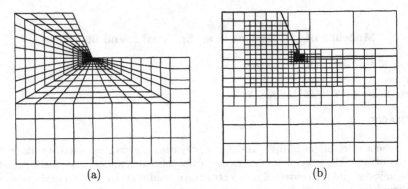

Figure 1: (a) Unstructured finite element mesh, and (b) resulting structured grid.

MODELING DISLOCATION DYNAMICS

Dislocations are moved by the forces acting on them. Once the force on a dislocation point has been found, one can move this point according to the dynamics that relates the dislocation velocity to the force.

The forces on a dislocation can be loosely categorized into four types: (1) the dislocation "line-tension" self-interaction, (2) dislocation-dislocation interactions, (3) the applied stress arising from thermal mismatch, lattice mismatch or other extrinsic mechanisms, and (4) the image forces needed to satisfy the boundary conditions at surfaces and interfaces. The first two forces arise from the dislocations themselves, and therefore change as the dislocations move. They are calculated by integrating over all dislocations present using the well-known Peach-Koehler formalism [2]. Although the theory of dislocations has been well developed for many years, the numerically intensive nature of such calculations has caused its application to lag far behind experiment. With the advent of powerful computers, however, it has become feasible to simulate realistic dynamics in a reasonable time. Thus the recent decade has seen intensive research in this area, resulting in several research codes based on the Peach-Koehler formalism [3-6]. In this paper we use the PARANOID code, where the reader is referred to [5] for the method and implementation.

There remain the issues of the applied stress and the image corrections. We discuss these separately before proceeding to the experimental description and the results.

Applied Force

The applied force, item (3) above, can be computed by the Peach-Koehler formula once the stress field is known. The determination of this field is usually a more or less difficult boundary-value problem in mechanics, and can be attacked using numerical techniques such as finite element or boundary element methods, which can deal with the complicated geometries that are characteristic of microelectronic devices. Many special-purpose commercial codes have been developed for this kind of problem. In the dislocated island the stress field due to lattice mismatch can be calculated using the finite element mechanical modeling package ABAQUS [7], as described in the next section.

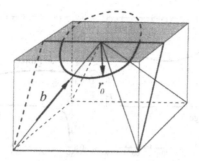

Figure 2: Semi-circular dislocation lying on a (111) plane and terminating on a (001) free surface.

Often, stress fields of extrinsic origin are highly non-uniform in space. As a dislocation moves, it therefore is subject to rapidly varying forces. This causes a practical difficulty, since the finite element method produces large, *unstructured* tables. That is, to move the dislocation one would need to search such a table for every dislocation point at every time step. This turns out to be very time consuming when thousands of dislocation points move over hundreds of thousands of time steps. To remedy this problem we first covert the results of the finite element calculation into a *structured* stress table.

Figure 1a shows a region discretized for the finite element method, and Fig. 1b is the structured grid for the same region. Note that in the area of refined mesh the structured grid is also refined in order to preserve the accuracy of the finite element calculations. Starting with a large box which encompasses the original region, the space is subdivided into identical smaller sub-boxes. The need for further refinement is then determined based on the finite element mesh and the stress gradient. If further refinement is required, another level of sub-boxes is created and is registered in a linked list which saves the lower-left corner coordinates of each sub-box. If further refinement is not required, the stress values at the corners of the box are calculated from the finite element results by interpolation and extrapolation. The refinement process continues down to as many refinement levels as are needed to maintain the spatial resolution of the finite element results.

With the structured table, the dislocation-motion stress search becomes very simple. From the location of the dislocation point one finds the index of a sub-box by comparing its coordinate with the lower-left corner of the current box. The linked list then tells one if there is a box on the next level which contains the point. The process continues, and one quickly arrives at the smallest box that contains the point. The stresses at the corners of the smallest box are then used to find the stress at the dislocation point by interpolation. Our experience shows that four or five levels of refinement usually suffice to convert a finite element table involving several tens of thousands of elements. This approach expedites the stress search by a factor of hundreds. Although illustrated for two dimension in Fig.1a, this conversion is readily implemented in three dimensions, where the computational advantages are even greater.

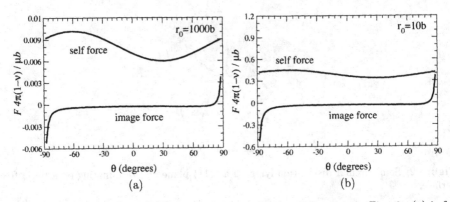

Figure 3: Image force and self force on the semi-circular dislocation in Fig. 2. (a) is for radius of 1000 times Burgers b, and (b) for 10 times b. $\theta = -90°$ and $90°$ correspond to the surface points. In the force normalization, μ is the shear modulus and ν is the Poisson ratio.

Image Force

Image forces arise from the interaction of dislocations with boundaries and interfaces. For simple dislocation configurations such as straight lines, closed-form solutions may be found scattered throughout the literature [8–10]. However, no analytical solutions exist for more general configurations of dislocations and boundaries. Even a numerical solution is often difficult. In general, for a given configuration, one has to solve a boundary value problem in order to satisfy the boundary conditions at surfaces and interfaces. For a case involving moving dislocations, this problem has to be solved over again every few time steps – an entirely impractical approach when many time steps are involved.

In microelectronic devices the surface to volume ratio increases with shrinking device size, leading to the natural expectation that image effects play a very important role. However, although image forces are generally very important for 2D problems and other situations involving straight dislocation lines, they are usually much less important in situations where the dislocations are strongly curved. Since curvature forces also increase as the scale becomes smaller, it is important to look at this issue more carefully. We have done this by examining several dislocation configurations for which the image forces can be computed with a high level of accuracy. On the basis of these calculations we conclude that the image forces in the dislocated islands constitute a relatively small effect, and can be neglected as a starting approximation.

To make this estimate we consider only a planar surface, but the qualitative conclusion will obviously hold for more complicated surfaces. A planar surface is used because the image fields can then be obtained by simple superposition using the Boussinesq-Cerruti formalism, rather than by solving a complicated boundary value problem. Our focus is on the interaction of curved dislocations with the planar surface, including the situation where dislocations actually terminate on a free surface.

Figure 2 shows a free planar surface with a semi-circular dislocation that ends on it and is extended to a closed circular loop. Once the loop is closed, the traction it exerts on the

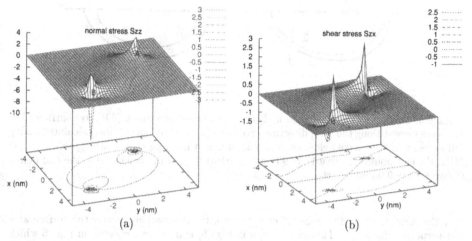

Figure 4: The surface traction due to the closed dislocation loop in Fig. 2. (a) is for normal stress, and (b) is for one of two shear stresses. The stresses are normalized by $\mu/4\pi(1-\nu)$.

surface can be evaluated using the Peach-Koehler formalism. The traction can be canceled utilizing known point force solutions for a free surface, i.e. the Boussinesq solution [11] for the normal force and the Cerruti solution [12] for the tangential force. The image force is then computed by inserting the resulting Boussinesq-Cerruti stress field into the Peach-Koehler equation. For details on image force calculations and regularization, the reader is referred to a forthcoming paper by Liu and Schwarz [13].

Figure 3 shows the calculated image force for two different loop sizes. For comparison, the self-force is also plotted. It is seen that for a large portion of the dislocation loop the image forces acting on it are close to zero — in particular, they are negligibly small compared to the self-forces. Only close to the surface points do the image forces become important. This behavior can be readily understood by examining the surface traction exerted by the closed loop. Figure 4 shows the traction for a loop of radius ten times the length of the Burgers vector. Even for a loop this small, the surface tractions are large only where the end points touch the surface, and decrease rapidly away from these regions. Note also the different scales on Fig. 3a and Fig. 3b. It is true that the smaller the dislocation loop size, the larger the image force, but the same holds for the self-force, and their *relative* importance changes only logarithmically.

We now discuss the dynamical behavior of such dislocation semi-loops under planar biaxial straining. As a particularly sensitive test, we examine the stationary loop configuration, below which a loop shrinks and beyond which it grows without limit. For comparison, we carry out two kinds of calculations, one with image forces *fully and accurately* included and the other with the line-surface interaction treated very approximately as described in [5]. It may be seen from Fig. 5 that, although there are some differences between the two calculations, it is a reasonably good approximation to neglect image forces. Furthermore, we conclude that the effects of the image forces are almost entirely confined to the region very

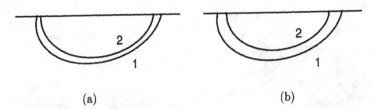

Figure 5: Stationary loops lying on a (111) plane and intersecting a (001) free surface. The loops are viewed along the [110] direction. For both (a) and (b), curve 1 was calculated taking full account of the image fields; curve 2 was obtained ignoring them. (a) $\sigma_{xx} = \sigma_{yy} = 105$ MPa; the end points of curve 1 are 954 nm apart; (b) $\sigma_{xx} = \sigma_{yy} = 5.06$ GPa; the end points of curve 1 are 9.05 nm apart.

near the surface, where the main effect is to force the dislocation to enter the surface at a non-perpendicular angle. This can be seen in Fig. 5, and in greater detail in Fig. 6 which shows some actual dynamical sequences produced by the PARANOID code supplemented by real-time evaluation of the image fields.

To summarize this section, we conclude that, although each case must be examined on its merits, the hard-to-compute image forces can be neglected for many problems, no matter what the scale. For such cases, applied and curvature forces are dominant, and the neglect of image forces does not introduce errors much greater than the 10-20% range already inherent in continuum dislocation dynamics. Furthermore, we find that such image corrections as exist are dominated by the singular interactions which occur when the dislocation enters the surface, and which cannot be treated by finite-element methods, unless these are resolved down to the scale of the Burgers vector.

EXPERIMENT

Islands of CoSi$_2$ grown on Si(111) were chosen as a test system for our simulations. This system has been studied in detail due to a long-standing interest in growing uniform epitaxial CoSi$_2$ films on Si for electronic applications [14–16], and the crystallography and growth mode of the silicide on Si(111) are well understood. Deposition of Co at high temperatures (above about 800°C) forms CoSi$_2$ directly without initial growth of sub-stoichiometric phases. The CoSi$_2$ grows epitaxially on the Si(111) surface with a B-type orientation, such that the two materials have a common (111) axis, but the CoSi$_2$ is rotated 180° [14,17–20]. However, a small lattice mismatch (around 0.6 % at the growth temperature) causes the CoSi$_2$ to grow as islands [17,20] rather than as a uniform film. These sub-micron sized islands are triangular or hexagonal and are flat-topped and bounded by (111) facets. In common with island growth in strained semiconductor systems such as Ge/Si or InGaAs/GaAs, smaller islands are pseudomorphically strained, while larger ones relieve the strain by the introduction of misfit dislocations. The dislocations remain at or near the CoSi$_2$/Si(111) interface [20], leading to a well-defined geometrical structure for comparison with simulations.

Transmission electron microscopy (TEM) was used to characterize island shapes and sizes and to observe the configuration of dislocations. To record island growth and dislocation motion, and to avoid any effects of a surface oxide, we formed the islands *in situ* in an ultra-

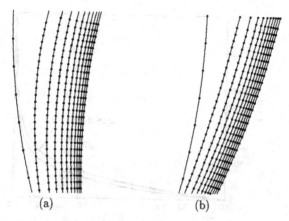

Figure 6: Motion of a dislocation loop for the points close to the surface when image forces are fully included. (a) is for the left end point and (b) is for the right. As time increases from left to right, the dislocation at both ends approaches the same angle with the surface.

high vacuum TEM. The microscope used, a Hitachi UHV H-9000, has a base pressure of 2×10^{-10} Torr and was modified to allow deposition to take place during observation [22,23]. Silicon specimens were chemically etched to a thickness of several microns, and then cleaned under UHV by heating to 1250°C. Final thinning to electron transparency was carried out by etching with oxygen at about 850°C and 10^{-6} Torr. Cobalt was deposited using an electron beam evaporator mounted just above the microscope objective lens [24,25]. During deposition, island growth was recorded at video rate using a CCD camera with an image intensifier. Real time observations of the growth of individual islands allowed us to monitor the introduction of the first dislocations and the subsequent evolution of the dislocation networks shown in Fig. 8.

RESULTS

In the island problem we need to find the stress field due to lattice mismatch between the island and the substrate. This was obtained using the commercial finite element package ABAQUS [7]. The data on the silicide mechanical properties were taken from Maex and van Rossum [26]. The island shape was simplified to that of a truncated cone with a flat top, where the sidewall of the cone forms an angle of 70.5° (the angle between {111} glide planes) with the base. As a result of this simplification, the island on its substrate becomes an axisymmetric problem. Axisymmetric eight-node isoparametric elements were used for discretization. The mesh was refined at the island edge to resolve the divergent stresses there, and 1/4-point elements were focused at the edge tip. The substrate was taken to be at least ten times the largest dimension of the island. The in-planar displacement of the axisymmetric axis was constrained, and the lowest point on the axis was fixed to remove rigid-body motion. Lattice mismatch was specified through the thermal mismatch capability in ABAQUS.

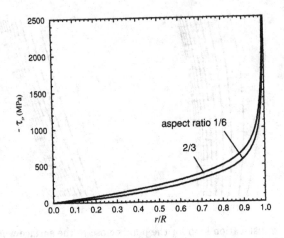

Figure 7: Shear stress at the island-substrate interface for two aspect ratios (height over radius), as a function of distance from island center.

The stress of most interest is the shear stress on the island and substrate interface. Figure 7 shows the calculated stress for two different aspect ratios. Because $CoSi_2$ has a smaller lattice constant than silicon, the island is under tension and the shear stress acting on the substrate points toward the island center. It is noted that the stress diverges at the island edge as expected. The stress close to the edge is consistent with the asymptotic solution which dictates that it varies as $d^{-0.415}$, where d is the distance from the edge. In addition, the figure shows that the stress increases with increasing aspect ratio.

With the stress obtained by finite element analysis, the structured stress table was constructed as described in previous section, and was incorporated into the dislocation dynamics code PARANOID to study dislocation behavior under a combination of self force, dislocation-dislocation forces, and applied force. In accord with the previous discussion, image forces were neglected. The interface is the {111} glide plane, and the possible Burgers vectors belong to $\frac{1}{2}\langle 110 \rangle$. These are the slip systems studied in this paper.

If a $\frac{1}{2}[110]$ dislocation is allowed to evolve into the island, the simulation shows that the dislocation spreads quickly along the island edge before it moves into the island. This is shown in Figure 8d in the form of the dislocation configuration at various times. The result agrees well with experimental observations (which do not resolve the immediate edge region), in that dislocations appear to come out suddenly from a large portion of the edge. Numerical simulation shows that this behavior does not depend on the nucleation location, but that the final orientation achieved by the dislocation is perpendicular to its Burgers vector. When all three possible Burgers vectors are allowed and multiple dislocations are introduced into the island, they evolve into stationary patterns, two of which are shown in Fig. 8e and Fig. 8f. The patterns are the result of the inherent island stress field, the self-interaction, and the interactions between dislocations as they cross. In comparing Figure 8e with 8b, one notices that even rather fine features are captured by the calculation, such as

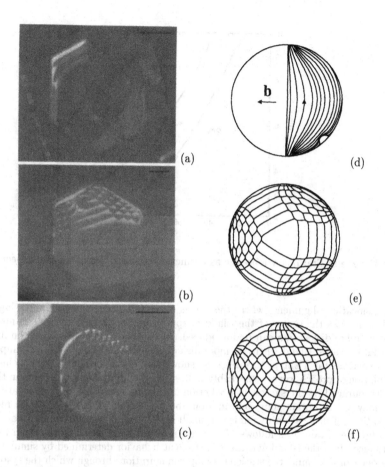

Figure 8: (111) plane-view of dislocations in CoSi$_2$ islands as observed by *in situ* transmission electron microscopy, and as simulated by 3D dislocation dynamics. The images were taken in a $\mathbf{g} = \langle 220 \rangle$ dark field condition for the Si substrate, showing the dislocations as lines of strong bright and dark contrast. The scale bar on each image represents 250 nm. (a) Initial entry of the dislocations, recorded at the growth temperature of 850°C. (b) Observed configuration after holding the specimen at 850°C for 30 min. (c) A second, smaller island, this time fully dislocated, recorded after cooling the system. The small circular features within each island are pinholes which form during growth. (d) Series of position snapshots showing the calculated evolution of a single $\mathbf{b} = \frac{1}{2}[110]$ dislocation loop introduced at the island edge. (e) Calculated transient configuration resulting from the introduction of six sets of dislocation loops, each containing the three possible $\frac{1}{2}[110]$ Burgers vectors for the (111) interface plane. (f) Fully relaxed pattern predicted by the simulations. Additional loops introduced into the island will not grow.

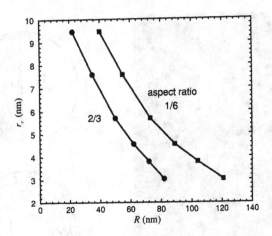

Figure 9: Stationary loop size as a function of island radius for two aspect ratios.

the dislocation alignments where they cross over each other, and the increasing spacing of dislocations as the center of the island is approached. The pattern in 8e is not saturated, that is, more dislocation can be introduced. In contrast, Figure 8f shows the limiting case of a fully relaxed island where no more dislocations can be added. Again, comparison with the experimental TEM picture Fig. 8c shows striking agreement. We conclude that the simulation approach works remarkably well, given the approximations made in the analysis.

Encouraged by the successful prediction of the observed stationary dislocation patterns, one may try to obtain some information about their nucleation. While we recognize the limitations of dislocation dynamics at small scales, useful qualitative information can nevertheless be obtained as follows. For a given island size, various dislocation loops can be introduced into the island and their subsequent behavior determined by simulation. In this way one can determine the stationary loop configuration through which the system must be activated if the island is to dislocate. Figure 9 shows the stationary loop size as a function of island ratio for two values of the aspect ratio. As the island grows larger, the stationary loop size decreases, meaning that it should be increasingly easier to nucleate a dislocation as the island grows. In addition, we observe that high aspect-ratio (thick) islands nucleate more readily than low aspect-ratio (thin) islands. These behaviors are in accord with experimental observations, and can be understood from plots of the shear stress shown in Figure 7. For a given aspect ratio, the stress is a universal function of r/R, where r is the distance from the island center, and R is the island radius. Thus, at a given distance $d = R - r$ from the edge, the stress increases as the island grows, making it easier to nucleate a dislocation. Similarly, thicker islands do not relax as much as thin islands, leaving them with higher internal fields. This again favors dislocation nucleation.

CONCLUSION

We have used the dislocation-simulation program PARANOID to calculate the behavior of dislocations in CoSi$_2$ islands grown by reactive epitaxy on Si(111) substrate. The stress fields in the islands are calculated using standard finite element modeling techniques, and are converted into a structured, multi-level and multi-grid stress table that is imported into the program to drive the dislocations. Single and multiple dislocations in the islands have been simulated, and the predicted patterns are strikingly similar to those observed experimentally. By looking at the growth of very small loops we also find that dislocation-loop nucleation becomes easier as the islands become larger, and that thick islands are dislocated at smaller size that thin ones. These results are also in good agreement with experimental observations. We conclude that current modeling techniques are sufficient to treat this type of problem at a useful level of accuracy.

Our study of dislocated islands shows that a combination of finite element analysis and dislocation dynamics, with the image force correction neglected, can in many instances predict dislocation behavior with good accuracy. The same machinery can be applied to other complicated microstructures. The eventual goal of predicting dislocation behavior during microelectronic device fabrication seems now to be within reach.

REFERENCES

[1] X. H. Liu, F. M. Ross, and K. W. Schwarz, *Phys. Rev. Lett.* **85**, 4088 (2000).
[2] J. P. Hirth and J. Lothe, *Theory of Dislocations*, (Krieger, Malabar, FL, 1992).
[3] B. DeVincre and L. P.Kubin, *Model. Simu. Mater. Sci. Eng.* **2**, 559 (1994).
[4] H. M. Zbib, M. Rhee, and J. P. Hirth, *Int. J. Mech. Sci.* **40**, 113 (1998).
[5] K. W. Schwarz, *J. Appl. Phys.* **85**, 108 (1999).
[6] N. M. Ghoniem and L. Z. Sun, *Phys. Rev.* **B60**, 128 (1999).
[7] *ABAQUS User's Manual Version 5.7* (Hibbitt, Karlsson, and Sorensen, Inc., Pawtucket, RI, 1997).
[8] J. D. Eshelby and A. N. Stroh, *Phil. Mag.* **42**, 1401 (1951).
[9] J. Dundurs, Elastic Interaction of Dislocations with Inhomogeneities, in *Mathematical Theory of Dislocations*, ed. T. Mura, (ASME, New York, NY, 1969) pp. 70-115.
[10] J. D. Eshelby, Boundary Problems, Chap. 3 in *Dislocations in Solids*, Vol. 1, ed. F.R.N. Nabarro, (North-Holland Publishing Company, 1979) pp. 169-221.
[11] J. Boussinesq, *Application des potentiels à l'étude de l'équilibre et du mouvement des solides élastiques,* (Paris: Gauthier-Villars, 1885) p. 45.
[12] V. Cerruti, *Roma, Acc. Lincei, Mem. fis. mat.* (1882).
[13] X. H. Liu and K. W. Schwarz, in preparation (2001).
[14] R. T. Tung, J. M. Poate, J. C. Bean, J. M. Gibson, and D. C. Jacobson, *Thin Solid Films* **93**, 77 (1982).
[15] H. von Kanel, *Mat. Sci. Rep.* **8**, 193 (1992).
[16] K. Maex, *Mat. Sci. and Eng.* **R11**, 53 (1993).
[17] J. M. Gibson, J. C. Bean, J. M. Poate, and R. T. Tung, *Thin Solid Films* **93**, 99 (1982).
[18] S. A. Chambers, S. B. Anderson, H. W. Chen, and J. H. Weaver, *Phys Rev.* **B34**, 913 (1986)
[19] J. M. Gibson, J. L. Batstone, and R. T. Tung, *Appl. Phys. Lett.* **51**, 45 (1987).

[20] C. W. T. Bulle-Lieuwma, D. E. W. Vandenhoudt, J. Henz, N. Onda, and H. von Kanel, *J. Appl. Phys.* **73**, 3220 (1993).

[21] P. A. Bennett, S. A. Parikh, and D. G. Cahill, *J. Vac. Sci. Technol.* **A11**, 1680 (1993).

[22] M. Hammar, F. LeGoues, J. Tersoff, M. C. Reuter, and R. M. Tromp, *Surf. Sci.* **349**, 129 (1995).

[23] F. M. Ross, F. K. LeGoues, J. Tersoff, R. M. Tromp, and M. C. Reuter, *Microscopy Res. Tech.* **42**, 281 (1998).

[24] F. M. Ross, P. A. Bennett, R. M. Tromp, J. Tersoff, and M. C. Reuter, *Micron* **30**, 21 (1999)

[25] F. M. Ross, J. Tersoff, R. M. Tromp, M. C. Reuter, and P. A. Bennett, *J. Electron Microscopy* **48**, 1059 (1999).

[26] K. Maex and M. van Rossum, editors, *Properties of Metal Silicides*, EMIS Data Reviews, (Inspec, London, 1995).

Mat. Res. Soc. Symp. Proc. Vol. 673 © 2001 Materials Research Society

MISFIT DISLOCATION INTRODUCTION DURING THE EPITAXIAL GROWTH OF InAs ISLANDS ON GaP

Vidyut Gopal[1], Alexander L. Vasiliev[2], and Eric P. Kvam
School of Materials Engineering, Purdue University, W. Lafayette, IN 47907
[1]: Currently at Applied Materials Inc., Santa Clara, CA 95051
[2]: Currently at the Institute of Materials Science, University of Connecticut, Storrs, CT 06269

ABSTRACT

The initial growth of InAs on 11% lattice mismatched GaP substrates by molecular beam epitaxy was investigated. High resolution transmission electron microscopy (HREM) images showed that the InAs grew in the form of three-dimensional islands of dissimilar sizes. Mismatch induced strain relief was effected by the direct introduction of (mostly) edge dislocations at the corners of the islands. An examination of HREM images of several islands revealed that the island aspect ratio decreased with the introduction of misfit dislocations. Strain relaxation in the smaller, relatively dislocation-free islands occurred by elastic deformation of InAs lattice planes, which was more effective far from the constrained island-substrate interface. As a result, these islands grew taller and narrower, with a gradient in the elastic strain energy. However, a higher aspect ratio resulted in a higher surface area – to – volume ratio, and increased the surface energy of the InAs islands. Consequently, there was a driving force for the reduction of the aspect ratio if an alternate avenue for strain relaxation existed. The alternate route was plastic deformation by the introduction of misfit dislocations. As the island grew, the strain at the island corners increased, and beyond a critical value, misfit dislocations were added. These dislocations relieved strain at the heterointerface, and promoted the islands to grow laterally, i.e., the aspect ratio decreased. Islands coalesced, and a continuous layer resulted by a nominal thickness of 3 nm. Thus, the morphology of InAs islands grown on GaP was determined by the balance between elastic and plastic deformation.

INTRODUCTION

The integration of semiconductor single crystals of different band gap offers great flexibility in the design of various electronic devices. Examples include devices fabricated using the GeSi/Si system and the AlGaAs/GaAs systems. Improvements in crystal growth techniques such as molecular beam epitaxy (MBE) and metal oxide chemical vapor deposition (MOCVD) allow precise control in film composition and thickness to be achieved, especially if the mismatch in lattice constant between the overlayer film and the substrate is not very large. However, several compound semiconductors are not lattice-matched to any of the commercially available substrates. One example is InAs. It has a narrow band gap (0.36 eV) and high electron mobility (33,000 cm^2/V-sec), which makes it suitable for far infra-red detecting applications, thermo-photo-voltaics, and fast transistors. Traditionally, InAs has been grown on GaAs with a lattice mismatch of ~ 7%. Here, we investigated the initial stages of the growth of InAs on GaP, a lattice mismatch of ~ 11%. The motivation for this choice is that GaP is closely lattice-matched to Si. The direct growth of InAs on Si was found to result in very poor quality films because of both the high lattice mismatch and a polar/non-polar interface [1]. Hence, the successful growth of InAs on GaP could be the first step in an integrated InAs/GaP/Si technology.

EXPERIMENT

InAs layers were grown on commercially obtained (001) GaP substrates by solid-source MBE using a Varian GEN-II system. The substrates were thermally cleaned at 710°C (as measured by a thermocouple) under P_2 overpressure. A 20 period superlattice consisting of 5 nm alternating layers of GaP and AlP was grown to prevent the outdiffusion of impurities from the substrate. This was followed by the growth of a 200 nm buffer layer of undoped GaP at 660°C. The substrate temperature was lowered to 350°C for the growth of 3 monolayers (nominally) of InAs. Transmission electron microscopy (TEM) was used to study the dislocation microstructure, including both high resolution TEM (HREM) of cross-sectional specimens and dark field imaging of plan view specimens.

RESULTS AND DISCUSSION

Cross-section HREM images and reflection high-energy electron diffraction (RHEED) patterns show that InAs grows epitaxially on GaP (001) in Volmer-Weber or island mode. The 11% lattice mismatch strain is relieved by a two-dimensional network of (mostly) edge misfit dislocations that run along the [110] and [1$\bar{1}$0] directions, with an inter-dislocation spacing of approximately 4 nm. The mechanisms of strain relaxation that result in this dislocation microstructure are discussed below.

HREM images provide evidence for two distinct dislocation sources during InAs growth:
1) Formation of edge misfit dislocations at the edges of islands, followed by lateral growth of the islands past the dislocations. Threading dislocation segments are not necessarily formed in this process.
2) Nucleation of dislocation loops at steps on the surface of an island and their subsequent glide toward the interface resulting in a 60° misfit dislocation and two threading dislocation segments.

Due to the 11% lattice mismatch, the driving force for strain relaxation was very high during initial growth, and sessile dislocations were nucleated directly at island edges from the outset. This is unlike lower mismatch systems such as Ge/Si where some coherent growth was found to occur before dislocation introduction commenced. Figure 1 is a HREM image of a small island that clearly has one 90° misfit dislocation, and the (bright) strain contrast at the two edges of the island (pointed out with arrows in the figure) might be indicative of incipient dislocation introduction at these edges. Once the islands coalesced, the energy barrier to direct introduction of edge misfit dislocations became very large. Consequently, the latter process – glissile misfit dislocation introduction – became the dominant mechanism as the InAs film grew thicker.

The equilibrium shape of an island in a non-wetting (Volmer-Weber) epitaxial system depends on the interfacial energy balance. Consider an island of isotropic material A deposited on a substrate B as shown in figure 2. The island, shown here as a spherical cap, is characterized by its contact angle, θ. The specific interfacial energies are γ_{sv}, γ_{fs} and γ_{vf}, where the subscripts s, f, and v represent substrate, film and vapor respectively. A one-dimensional force balance of the interfacial tensions yields Young's equation:

$$\gamma_{sv} = \gamma_{fs} + \gamma_{vf} Cos\theta$$

However, this assumes that the materials constituting the island and the substrate are lattice matched, and that there is no residual strain in the island or substrate that could alter the energy balance. If this equation were to govern the system, ignoring such effects as faceting of the island surface, the equilibrium contact angle should remain the same as growth proceeds. Such growth is termed self-similar. This was not the case for InAs growth on GaP.

Figure 1: HREM image of a small InAs island with one 90° misfit dislocation.

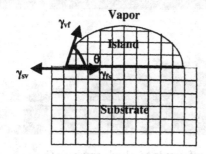

Figure 2: Schematic figure of surface and interface energies for a solid-solid epitaxial system.

Figures 3a – c are HREM micrographs of different InAs islands. Each of these images, as well as figure 1 came from the same sample, with a nominal thickness of 4 monolayers (i.e., the thickness that would result if growth were uniform). The InAs islands grew in a wide variety of shapes and sizes, with different strain and misfit dislocation content. This was fortuitous, since it offset the need to grow many samples, each with a slightly different thickness, to obtain snapshots of the growth process. By imaging different regions of the same sample, information linking the morphology of islands to the strain and dislocation content was collected.

The aspect ratio (ratio of height to width) of an island scales directly with the contact angle ($\theta = \mathrm{Cos}^{-1}[h/w]$). Figure 4 plots the aspect ratio versus the number of misfit dislocations for several islands. (The data for this plot were obtained from HREM images of cross-section specimens, which are images of two-dimensional sections through the islands. Only the dislocations running in one of the two in-plane <110> directions were visible, and hence only these were counted. "Height" refers to the peak height of an island in a HREM image and "width" refers to basal width in the same HREM image) It is clear that the growth is not self-similar. The aspect ratio decreases as the dislocation content increases, and more strain is relieved. Thus, dislocation introduction leads to enhanced lateral growth of islands at the expense of vertical growth – flattening of the islands.

a)

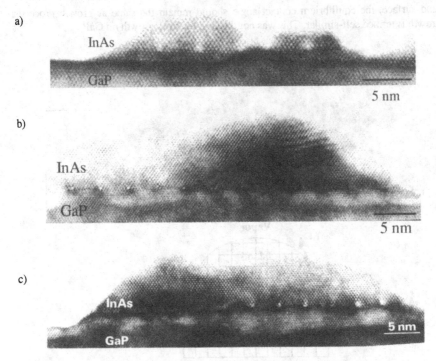

Figure 3: HREM micrographs of different InAs islands grown on GaP.

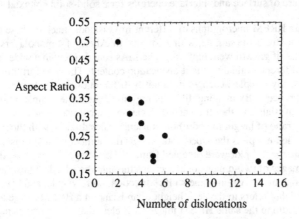

Figure 4: Variation of island aspect ratio with misfit dislocation content

This phenomenon can be understood on the basis of energy minimization. The free energy of the island-substrate system consists of the elastic strain energies in the island and the underlying substrate, the surface and interfacial energies of the film and substrate materials, and the energy of the dislocation network at the interface. If the island material is lattice matched to the substrate there is no elastic strain in the system and the island adopts the aspect ratio predicted by Young's equation. This condition is shown schematically in figure 2. If the island were to grow coherently (i.e., without misfit dislocation introduction), then the balance between strain energy and surface energy would determine its equilibrium shape. Since the island is constrained by the substrate only near the interface, a taller and narrower island could undergo elastic deformation more efficiently to relieve strain. Under compressive strain, as for InAs on GaP, the island tends to expand along the interface but is constrained by the substrate. It responds instead by bending, so that the interfacial plane becomes curved. This is shown schematically in figure 5a. Since the constraint is only at the interface, the magnitude of the strain diminishes rapidly with distance from the substrate, i.e., the strain distribution is spatially non-uniform. But such elastic strain relaxation occurs at the cost of increased island surface energy. The island would tend to reduce its surface energy (and hence its contact angle) by growing laterally, if an alternate (lower energy) avenue for strain relaxation were available. Plastic deformation by misfit dislocation introduction is the necessary alternate means of strain relief. This condition is shown schematically in figure 5b. FEM studies performed by Johnson and Freund on the mechanics of a coherently strained island, as well as an island with one dislocation, predicted that the aspect ratio would decrease upon dislocation introduction [2].

Edge (90°) dislocations (and sometimes 60° dislocations as well) were directly introduced at the edge of a growing island, and due to the local strain relief, the island grew laterally past the dislocation until strain energy built up again at the edge. When the mismatch strain induced force exceeded a critical value, a new dislocation was added at the edge and the process repeated. Such a mechanism is also supported by the work of LeGoues et al., who recorded in-situ TEM images of the growth of Ge islands on Si, and observed a cyclic growth pattern [3]. Islands grew vertically until a dislocation was introduced at the edge, followed by a spurt of lateral growth.

a) 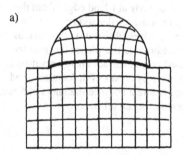   b)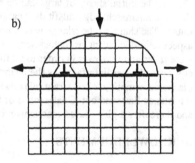

Figure 5: Schematic diagrams of island morphology in three different conditions: a) Coherent (i.e., elastically, without dislocations) strained; b) Strain relaxed by misfit dislocations.

This cyclic dislocation assisted island growth process cannot continue indefinitely. As the islands spread laterally, they soon meet and coalesce. After the formation of a continuous layer, direct introduction of Lomer dislocations ceased to be operative. A glide-based mechanism was then necessary to complete strain relaxation. The details of this mechanism have been discussed elsewhere by us, and will not be dwelt upon here [4].

Additional information about island morphology was obtained through plan view imaging. Figure 6 is a $\bar{2}20$ dark field plan view micrograph of InAs islands. The islands have a wide size distribution. Moire fringes running parallel to [110], which result from the interference of electron beams diffracted by crystals of different lattice spacing, are clearly visible. They indicate the presence of strain relaxed InAs islands. The smaller islands appear equiaxed. However, as the islands grew they developed facets along the [100], [010], [110], or [$\bar{1}$10] directions and their basal shape was either square or triangular.

Figure 6: 220 dark field plan view TEM micrograph of InAs islands grown on GaP.

CONCLUSIONS

The initial stages of large-lattice mismatched InAs/GaP epitaxy were investigated. Due to the 11% mismatch, edge misfit dislocations were introduced directly at island edges from the outset. The shape of the islands was strongly influenced by the misfit dislocation content, with aspect ratio decreasing as dislocations were added. Smaller, relatively dislocation free islands were taller and narrower in order to facilitate elastic deformation of lattice planes far from the constrained heterointerface. As the island grew, the increased cost of surface energy resulted in plastic deformation – i.e., strain relaxation at the heterointerface by means of dislocations, and the islands flattened out. Coalescence of the islands resulted before a nominal thickness of 5 nm, and thereafter a glide-based mechanism continued the process of strain relaxation.

ACKNOWLEDGEMENTS

The authors would like to acknowledge Prof. J.M. Woodall and Dr. E.-H. Chen of Yale University for providing MBE grown samples. This research was supported in part by the National Science Foundation through grant DMR-9400415.

REFERENCES

1. Desikan A., M.S. Thesis, Purdue University (1999).
2. Johnson, H.T., and Freund, L.B., *J. Appl. Phys.*, 81, 6081, (1997).
3. LeGoues, F.K, et al., *Phys. Rev. Lett.*, 73, 300, (1994).
4. Gopal V., Vasiliev A.L., and Kvam E.P., accepted to appear in *Philos Mag. A* (2001).

Mat. Res. Soc. Symp. Proc. Vol. 673 © 2001 Materials Research Society

X-Ray Diffuse Scattering from Misfit Dislocation at Buried Interface

Kaile Li, Paul F. Miceli, Christian Lavoie,[1] Tom Tiedje,[1] and Karen L. Kavanagh[2]
Dept. Physics and Astronomy, U. Missouri-Columbia, Columbia, MO, 65211 USA
[1]Dept. Physics and Astronomy, U. British Columbia, Vancouver, BC, Canada V6T 1Z1
[2]Dept. Physics, Simon Fraser University, Burnaby, BC, Canada V5A 1S6.

ABSTRACT

Motivated by x-ray scattering experiments on heteroepitaxially grown thin films, we present model calculations of the diffuse x-ray scattering arising from misfit dislocations. The model is based on the elastic displacements from dislocations whose positions are spatially uncorrelated. These numerical results give support to a phenomenological model [Phys. Rev. B **51**, 5506 (1995)] that predicts the scaling of diffuse scattering intensity with perpendicular wavevector, Q_z. At low Q_z the diffuse width scales inversely with the defect size, which is given by the film thickness due to the effect of the elastic image field, whereas at high Q_z the diffuse width is mosaic-like, scaling with Q_z. New experimental results for $In_xGa_{1-x}As/GaAs$ are also presented and compared to the model. The calculations are in good agreement with these experiments, as well as other measurements in the literature for high and low dislocation density.

INTRODUCTION

It is well established that x-ray "rocking curves" from a bulk crystal containing dislocations have a "mosaic line shape" where the angular peak width is the same at all orders of Bragg reflection [1]. Recently, however, it has been shown that the scattering from misfit dislocations in thin epitaxial films exhibits different behavior due to the constraints on displacement fluctuations arising from the flat substrate: a scan taken transverse (in the direction along the surface) to a Bragg reflection can have "delta" (resolution-limited) and diffuse components of scattering. This effect has been addressed by several groups [2-4], although there exists only a limited amount of comparison between theory and experiment.

Based on phenomenological grounds it was suggested that there are two scattering regimes [4,5]: low "effective disorder" where the diffuse scattering is weak and dominated by the defect size whereas for high disorder the diffuse scattering is strong and dominated by the rotational effects of local displacements, thereby leading to mosaic-like features. This approach neglects the correlated scattering that might occur between dislocations, although, this should be a good approximation for either very low or very high dislocation densities [5].

In the present paper, we summarize some results from our recent efforts [6] to verify these phenomenological-based predictions using elasticity theory. Essentially, the elastic calculations show that there are, indeed, two scattering regimes as predicted. Moreover, in the weak diffuse scattering regime, we find that the width of the scattering is characteristic of the defect size, which turns out to be on the order of the film thickness. This length-scale is determined by the elastic dipole field that is generated by the dislocation and its image due to the free surface. We also present preliminary results for $In_xGa_{1-x}As/GaAs$ thin films where we show that the model fits the measured diffuse scattering over nearly three decades of intensity for the case of high dislocation density.

ELASTIC MODEL CALCULATIONS

The model assumes an elastically homogeneous medium where the film and substrate have the same elastic constants and the misfit dislocations are placed a distance below the free surface by an amount equal to the film thickness, d. This captures the dominant effect of the dislocation image field arising from the free surface. The elastic displacement fields, $\mathbf{u}$, for this geometry are given by Kaganer et. al. [3] who have also presented a x-ray scattering model. However, we have simplified the x-ray scattering problem [6] by recognizing that the spatial variation of $\mathbf{u}$ is negligible along the direction perpendicular to the surface. The relative scattered intensity shows "delta" (resolution-limited) and diffuse components according to:

$$\frac{I}{I_0} = \int_{-\infty}^{\infty} e^{iQ_x x} e^{-\rho d\gamma(x)} dx = 2\pi e^{-\rho d\gamma(\infty)} \delta(Q_x) + 2\int_0^{\infty} \text{Cos}(Q_x x)\{e^{-\rho d\gamma(x)} - e^{-\rho d\gamma(\infty)}\} dx$$

where,

$$\gamma(x) = \frac{2}{d}\int_0^{\infty} \text{Sin}^2[\frac{Q_z \Delta u_z(x, x_0, z)}{2}] dx_0 \quad,$$

and

$$\Delta u_z(x, x_0, z) = u_z(x + x_0, z) - u_z(x_0, z) \quad.$$

Here, the x and z directions are taken as parallel and perpendicular to the surface, respectively, and the corresponding wavevector components are Q_x and Q_z. The dislocation density is ρ. Because both experiment [4] and elastic calculations [5,6] indicate that the variation in $\mathbf{u}(x,z)$ with z is small, we evaluate $\mathbf{u}(x,z)$ at a single value of z taken at the midpoint of the film.

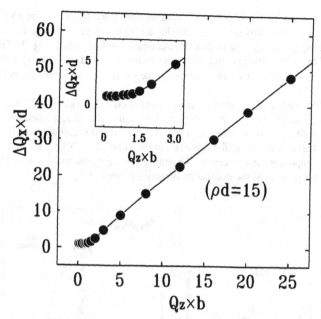

Figure 1. The width of the diffuse scattering parallel to the surface, ΔQ_x, plotted versus the perpendicular wavevector, Q_z, exhibits two scattering regimes.

The calculated results for the case where the Burgers vector, b, is oriented along x are given in Fig. 1, which shows the width of the diffuse scattering parallel to the surface, ΔQ_x, plotted versus Q_z. Two regimes of scattering are clearly seen. At low Q_z the diffuse scattering width, ΔQ_x, saturates to essentially the inverse film thickness, which is the lateral decay length scale for the dislocation displacement field in the presence of the free-surface boundary condition. It is found that the delta component is strong, but attenuating with Q_z, and the diffuse scattering is weak. On the other hand, a second scattering regime appears at higher Q_z where ΔQ_x is observed to scale linearly with Q_z, suggesting that the (mosaic-like) rotational character of the local displacement field dominates the diffuse scattering. Here, the delta component is very weak or absent. Results similar to Fig. 1 are observed [6] for the case where b is oriented along z. We find that the transition "knee" between the two regimes moves to higher Q_z as the dislocation density decreases. Thus, it might be that samples with small ρ exhibit only the first regime of scattering whereas samples with large ρ exhibit the second, mosaic-like, regime. This behavior was observed experimentally for ErAs/GaAs [4].

EXPERIMENT

In order to obtain further insight on the diffuse scattering behavior, measurements were performed on $In_{0.18}Ga_{0.82}As/GaAs(001)$ and some preliminary results are reported here. The films were grown by molecular beam epitaxy (MBE) and the details are given in refs. [7] and [8]. The x-ray measurements utilized MoK$_{\alpha2}$ (0.70926A) radiation from a line focus rotating anode source, and a Ge(111) monochromater provided a beam having a very low angular divergence, ≤0.003°. The reflection scattering geometry used in these measurements is illustrated in Fig. 2. Transverse scans were obtained by holding the perpendicular wavevector, Q_z, fixed at the Bragg reflection while rotating the angle, θ. The wavevector parallel to the surface is then simply given by $Q_x = \Delta\theta \times Q_z$.

Increasing the film thickness generally leads to an increased dislocation density and this is demonstrated in Fig. 3, which shows transverse scans across the (004) Bragg reflection for two samples with different film thickness. The 33nm film shows no diffuse scattering and the peak is resolution limited, whereas the 250nm film exhibits a broad diffuse line shape without a measurable "delta" component. A decrease of the delta component concomitant with an increase in diffuse scattering is expected as the dislocation density increases [3,4].

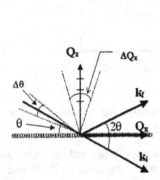

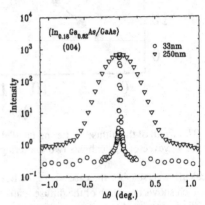

Figure 2. X-ray scattering geometry.

Figure 3. Transverse scans through the (004) reflection of two samples.

We now turn to a discussion of the diffuse scattering from the 250nm sample. Figure 4 shows the intensity, normalized to the peak value and presented on a logarithmic scale, measured for transverse scans across the (002), (004) and (008) Bragg reflections. The higher "background" signal for the (002) and (008) reflections is simply due to the fact that we are plotting the normalized intensity and that these reflections are weaker than the (004). As can be seen, the angular width of the diffuse scattering is the same at each reflection. This fact is further illustrated by the inset, which shows the full-width-at-half-maximum (FWHM) of the peak width plotted versus perpendicular wavevector. The observed linear behavior is just the mosaic-like scattering regime predicted in Fig. 1 for large wavevector or large dislocation density. Moreover, in addition to observing the width scaling that is determined on a linear intensity scale, we also show that our model fits the data on a logarithmic intensity scale: the solid curve fits the (004) reflection data over nearly 3 decades of intensity.

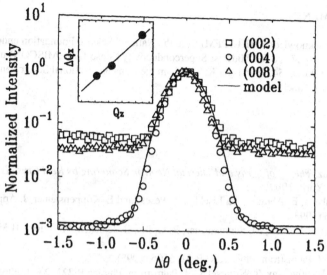

Figure 4. Transverse scans through Bragg reflections of the 250 nm film. Inset shows the FWHM of the diffuse scattering plotted versus the perpendicular wavevector.

DISSCUSSION AND CONCLUSION

There is now ample experimental evidence for the two scattering regimes depicted in Fig. 1. The defect size-limited scaling (ΔQ_x = constant) has been seen in ErAs/GaAs [4] and Nb/Al$_2$O$_3$ [9]. The mosaic-like scaling has been seen in ErAs/GaAs [4], Nb/Al$_2$O$_3$ [10] and here, in In$_x$Ga$_{1-x}$As/GaAs. Moreover, the evolution of the delta as well as diffuse scattering observed in experiments [3,4] is consistent with the model we have presented. However, what has not yet been clearly established experimentally is the crossover between the two scaling regimes within the *same* sample. The difficulty in observing this could be due, in part, to the "knee" in Fig. 1 that shifts with dislocation density, thereby pushing samples with small ρ toward the low-Q regime and samples with large ρ toward the high-Q regime. Furthermore, the crossover region (i.e. close to the "knee" of Fig. 1) is anticipated [5] to show correlation effects where the diffuse scattering can exhibit lobes.

In conclusion, elastic calculations have been used to show that two regimes are expected for scattering from independent misfit dislocations. At low-Q_z, weak diffuse scattering having a width that scales with the inverse film thickness is observed whereas at high-Q_z the diffuse scattering is strong (with a weak delta component) and it exhibits mosaic-like scaling. A number of experimental observations support these results.

ACKNOWLEGEMENTS

Support is gratefully acknowledged (KL, PFM) from the National Science Foundation under contract no. DMR-9623827 and the Midwest Superconductivity Consortium (MISCON) under DOE grant DE-FG02-90ER45427; and (KLK, TT) from the Canadian National Science and Engineering Research Council.

REFERENCES

1. M. A. Krivoglaz, *Theory of X-ray and Thermal Neutron Scattering by Real Crystals* (Plenum, New York, 1969).
2. V. Holý, J. Kubena, E. Abramof, K. Lischka, A. Pesek, and E. Koppensteiner, J. Appl. Phys. **74**, 1736 (1993).
3. V. M. Kaganer, R. Köhler , M. Schmidbauer, R. Opitz., B. Jenichen, Phys. Rev. B **55**, 1793 (1997).
4. P.F. Miceli, C. J. Palmstrøm, Phys. Rev. B **51**, 5506 (1995).
5. P.F. Miceli, J. Weatherwax, T. Krentsel, C. J. Palmstrøm, Physica B **221**, 230 (1996).
6. K. Li and P. F. Miceli, to be published.
7. C. Lavoie, T. Pinnington, T. Tiedje, R.S. Goldman, K. L. Kavanagh, J. L. Hutter, Appl. Phys. Lett. **67**, 3744 (1995).
8. K. L. Kavanagh , M. A Capano, and L. W. Hobbs, J. C. Barbour, P. M. J. Marée, W. Schaff, J. W. Mayer, D. Pettit, J. M. Woodall, J. A. Stroscio, and R. M. Feenstra, J. Appl. Phys. **64**, 4843 (1988).
9. A. Gibaud, D. F. McMorrow and P. P. Swaddling, J. Phys. Condens. Matter **7**, 2645 (1995).
10. P. M. Reimer, H. Zabel, C. P. Flynn and J.A. Dura . Journal of Crystal Growth **127**, 643 (1993). P. M. Reimer, private communication.

Dislocations and Deformation
in Epitaxial Layers

Mat. Res. Soc. Symp. Proc. Vol. 673 © 2001 Materials Research Society

Development of Cross-Hatch Morphology During Growth of Lattice Mismatched Layers

A. Maxwell Andrews, J.S. Speck, A.E. Romanov[1], M. Bobeth[2] and W. Pompe[2]
Materials Department, University of California, Santa Barbara
Santa Barbara, CA 93106-5050, U.S.A.
[1]A.F.Ioffe Physico-Technical Institute, Russian Academy of Sciences
St. Petersburg 194021, Russia
[2]Technical University of Dresden
Dresden 01609, Germany

ABSTRACT

An approach is developed for understanding the cross-hatch morphology in lattice mismatched heteroepitaxial film growth. It is demonstrated that both strain relaxation associated with misfit dislocation formation and subsequent step elimination (*e.g.* by step-flow growth) are responsible for the appearance of nanoscopic surface height undulations (0.1-10 nm) on a mesoscopic (~100 nm) lateral scale. The results of Monte Carlo simulations for dislocation-assisted strain relaxation and subsequent film growth predict the development of cross-hatch patterns with a characteristic surface undulation magnitude ~50 Å in an approximately 70% strain relaxed $In_{0.25}Ga_{0.75}As$ layers. The model is supported by atomic force microscopy (AFM) observations of cross-hatch morphology in the same composition samples grown well beyond the critical thickness for misfit dislocation generation.

INTRODUCTION

In lattice mismatched semiconductor systems, large undulations in the surface height profile appear as a characteristic cross-hatch pattern, which can be revealed by atomic force microscopy (AFM), scanning tunneling microscopy (STM), and by optical Nomarski microscopy [1-4]. Cross-hatch morphology is a common feature for low mismatch (<2%) systems, *i.e.* InGaAs on GaAs and SiGe on Si, which grow in a planar 2D layer-by-layer mode [4-6]. Figure 1 illustrates the most common crystallography for dislocation assisted strain relaxation during (001) epitaxial film growth of face centered cubic (fcc) materials. Relaxation occurs via threading dislocation (TD) motion and misfit dislocation (MD) formation on the inclined {111} glide planes [7]. The orthogonal <110> directions, along which the cross-hatch pattern develops, are the same directions as the intersections of the glide planes with the film surface.

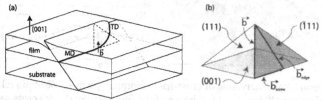

Figure 1. Dislocation geometry for a strained heteroepitaxial film with an inclined slip plane. (a) gliding threading dislocation segment with trailing misfit dislocation; (b) dislocation Burgers vector is decomposed into the $\vec{b}_{edge}$ and $\vec{b}_{screw}$ components.

The origin and evolution stages of cross-hatch morphology remain controversial and unresolved. One proposed mechanism for cross-hatch development relies on enhanced growth over strain relaxed regions [3,8-10]. The other mechanisms directly relate the height undulations with generation and glide of dislocation in the course of strain relaxation [2,6,11,12]. In this paper we will present a model for the onset of cross-hatch that incorporates both the strain relaxation in the film interior and the subsequent film growth.

EXPERIMENTAL OBSERVATION OF CROSS-HATCH

Epitaxial films of $In_xGa_{1-x}As$ were grown on GaAs (001) semi-insulating substrates in a Varian Gen II molecular beam epitaxy system. The films were grown at 520 °C with compositions of 15% In and 25% In, which correspond to 1.1 and 1.8% mismatch and a critical thickness approximately 100 and 50 Å, where h_c is the equilibrium critical thickness for misfit dislocation nucleation at film/substrate interface [7]. The film thicknesses h were 10, 20, 30, and 60 times h_c. The degree of relaxation and film composition were determined by {115} off-axis high-resolution x-ray diffraction (XRD) measurements [13]. Atomic force microscopy was used to determine the surface height profile. All films were partially relaxed. Figure 2 presents an AFM image of the surface of a 1000 Å (20 h_c) $In_{0.25}Ga_{0.75}As$ film that is 70% strain relaxed. The experimentally observed cross-hatch patterns for III-V semiconductors exhibit an anisotropy in the initial relaxation between the two orthogonal <110> directions, which is measurable by XRD and visible in the cross-hatch pattern. For the films in this study, the extent of strain relaxation were found to differ by up to 20% in the orthogonal directions. The peak-to-valley amplitude in the observed films increases with relaxation of the initial film strain. The relaxation of approximately 5, 20, 50, 70, and 100% resulted in maximum cross-hatch amplitudes of approximately 15, 25, 45, 60, and 100 Å, respectively (results for ~ 70% relaxed film are shown in Fig. 2). It is important to note that the film relaxation increases with increasing film thickness.

MODELING OF CROSS-HATCH

Below we propose a model for the surface height profile developement. Figure 3 illustrates the general idea for the evolution of the film surface undulations. Starting from a fully coherent strained epitaxial film (Fig. 3a), the film partially relaxes, via dislocation nucleation and motion, forming surface steps (Fig. 3b). Steps are then eliminated by their lateral flow resulting in the surface profile shown in Fig. 3c. The described sequence of the events can be also

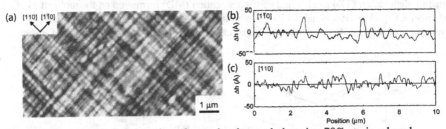

Figure 2. Experimental observation of cross-hatch morphology in ~70% strain relaxed $In_{0.25}Ga_{0.75}As$ film (h = 20 h_c). (a) AFM image of the film surface; (b), (c) cross sections of two orthogonal <110> directions in the AFM cross-hatch pattern.

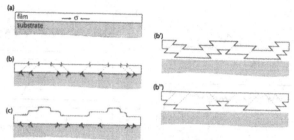

Figure 3. Schematic representation of the cross-hatch morphology development in strained films. (a) coherent compressively strained film; (b) strain relaxation in the film by dislocation glide resulting in the formation of surface steps and misfit dislocations; (c) final film state resulting from subsequent film growth via surface step elimination. The transition from (a) to (c) can be represented as Eshelby-like process where the unconstrained film plastically relaxes (b') then the steps are eliminated by subsequent growth (b'') and then the film is reattached to substrate (c).

in the surface profile shown in Fig. 3c. The described sequence of the events can be also described with the help of a hypothetical Eshelby-like procedure [14] where the strained film is removed from the substrate and relaxed by slip (Fig. 3b') [14]. The surface is then smoothed by film growth or lateral mass transport (Fig. 3b''). Finally the film is reattached to the substrate. The underlying slip steps can be directly affiliated to misfit dislocations at film substrate interface.

The details required to model the cross-hatch pattern can be separated into geometry, strain relaxation, and growth. The geometry for the (001) oriented surface with {111} slip systems is shown in Fig. 1. Only edge dislocations with the component of Burgers vector parallel to the film/substrate interface relieve the misfit strain in the film. That leaves two possible dislocation Burgers vectors per slip plane with both screw and edge components that create a surface step. The edge Burgers vector component $\vec{b}_{edge}$ can be decomposed again into the edge component parallel $\vec{b}_{\parallel}$ and perpendicular $\vec{b}_{\perp}$ to the surface. To calculate the surface displacement and stresses from the underlying misfit dislocations we use fully analytical elasticity solutions for edge dislocations with Burgers vectors oriented parallel or perpendicular to the surface [15]. For example, the component of stress $\sigma_{yy}^{b_{\perp}}$, resulting from a misfit dislocation with the Burgers vector perpendicular to the free surface is:

$$\sigma_{yy}^{b_\perp} = \frac{Gb_\perp}{2\pi(1-v)}\left(\frac{-x\left(x^2+3(y+h)^2\right)}{\left(x^2+(y+h)^2\right)^2} - \frac{x\left(x^2+3(y-h)^2\right)}{\left(x^2+(y-h)^2\right)^2} + \frac{2x\left(x^4+4x^2(y+h)^2-2x^2yh+3(y+h)^4+6yh(y+h)^2\right)}{\left(x^2+(y+h)^2\right)^3}\right).$$

(1)

The surface displacement expressions for perpendicular and parallel misfit dislocations Burgers vectors are given in Eqs. (2): x, y are spatial coordinates where the y-axis is normal to the surface, h is the film thickness and G and v are shear modulus and Poisson ratio of the material of the film.

$$u_y^{b_\parallel}\Big|_{y=0} = \frac{b_\parallel}{\pi}\frac{h^2}{x^2+h^2}, \quad u_y^{b_\perp}\Big|_{y=0} = \frac{b_\perp}{\pi}\left(-\frac{xh}{x^2+h^2}-\tan^{-1}\left[\frac{x}{h}\right]\right).$$

(2)

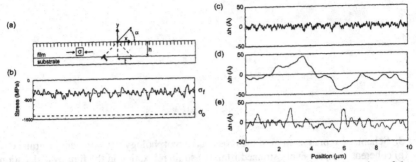

Figure 4. Modeling surface stresses and displacement. (a) schematic representation of the 1 D model; (b) resulting stress (σ_{xx}) at the surface after 70% strain relaxation; (c) surface profile created during film relaxation without step elimination; (d) surface profile after step elimination; (e) experimentally observed surface profile for strain relaxed $In_{0.25}Ga_{0.75}As$ film.

In our description, the surface step created during dislocation nucleation and motion was introduced by adding a Heaviside step function to the surface displacement caused by the subsurface misfit dislocation. This step is located at the intersection of the dislocation glide plane and the surface. To account for the surface profile after complete step elimination we use the displacement from Eqs. (2), which produces no slip step for the subsurface misfit dislocation.

A Monte Carlo algorithm was then used to simulate the misfit dislocation array generation and surface steps. Figure 4 illustrates the application of the algorithm for modeling a ~70% strain relaxed $In_{0.25}Ga_{0.75}As$ film. In this example, the so-called "left"-generated MD with Burgers vector components $\vec{b}_{\parallel} = -\dfrac{b}{2}\vec{i}$ and $\vec{b}_{\perp} = -\dfrac{b}{\sqrt{2}}\vec{j}$ is shown. Each possible surface nucleation site was included in a one-dimensional model, Fig. 4a, and the dislocation nucleation site was randomly selected. Then the glide plane was randomly selected and the resulting surface stresses and displacements were calculated. The nucleation of the next dislocation was modeled in the same way and the resulting total stress and displacements were calculated by linear superposition of surface displacement from the individual misfit dislocations and the related surface steps. The introduction of dislocations continued until a final value of relaxation was realized. The final stress profile at the surface (σ_{xx}), average film stress after relaxation (σ_f) and the original stress (σ_0) of the fully coherent film are shown in Fig. 4b. The surface height profile without film growth and lateral mass transport results is a locally rough film (Fig. 4c). Elimination of the surface steps (Fig. 4d) results in a locally smooth surface with large height amplitude undulations. The experimental height profile for the same film composition is shown in Fig. 4d.

RESULTS

For comparison with experimental data, the material properties of $In_xGa_{1-x}As$ on GaAs (001) substrate were used in the model. The results of the model do depend on material properties. The surface stress and surface height profiles were calculated for various extents of

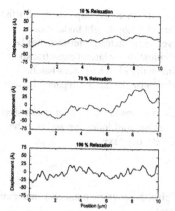

Figure 5. Surface displacement without slip steps as a function of film relaxation for a 1000 Å thick $In_{0.25}Ga_{0.75}As$ film

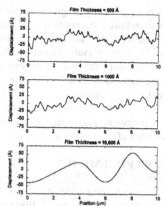

Figure 6. Surface displacement without slip steps as a function of thickness for a completely relaxed $In_{0.25}Ga_{0.75}As$ film

strain relaxation and for different film thickness and representative results are shown in Fig. 5 and 6. The results of the model share many important trends with the experimental data. Figure 5 illustrates that fully strained and only slightly relaxed films are significantly smoother than the more relaxed films. The average amplitude of the surface undulations and their apparent period both increase with increasing film relaxation. A film surface with only the slip steps is mesoscopically smooth (~100 nm length scale) but locally very rough. While the film surface with eliminated surface steps is locally smooth, it is a mesoscopically rough surface with large amplitude undulations.

The surface profiles are thickness dependent. Figure 6 shows the change in surface profile with varying film thickness. For very thin films, the displacement from misfit dislocations given by Eqs. (2) is laterally confined. Thus the surface appears to have many sharp transitions in the profile. As the film thickness increases and dislocation screening is more prevalent, the surface height transitions are more gradual. The surface of thicker films develops larger undulating amplitude and a locally smoother appearance.

The qualitative agreement of the experimentally observed surface cross-hatch with the modeled surface profile is good. The influence of thickness and strain relaxation on surface morphology is very similar. The quantitative discrepancies in the overall height amplitude and frequency of undulations that are seen in Fig. 4d and 4e can be attributed to many factors not currently included in the model. For example, stress induced surface diffusion is neglected. The incomplete elimination of steps due to surface diffusion would lead to a rougher surface than for the complete step elimination. The simulation algorithm applied here introduces only a single dislocation and slip step during each iteration, so multiple steps do not interact and do not form step bunchings. Real films can in addition have dislocation multiplication sources in the bulk.

CONCLUSIONS

A model for the evolution of the surface cross-hatch pattern has been presented. The model illustrates many important features of cross-hatch morphology development. Strain relaxed films are locally rough on an atomic scale and smooth on a mesocopic scale. Surface step flow is required to smooth out the local roughness and develop height undulations characteristic of cross-hatch. The experimentally observed cross-hatch pattern can not be explained by strain relaxation alone. Subsequent step elimination by lateral mass transport creates surface displacements on the order of the experimentally observed cross-hatch. Incomplete step elimination or step bunching could be the cause for the discrepancy between the model and the experimental.

This model will enable the use of cross-hatch observations to learn about misfit strain relaxation pathways, particularly dislocation sources, blocking, multiplication and possible routes to more efficient relaxation in lattice mismatched heteroepitaxy.

ACKNOWLEDGEMENTS

This work was supported by DAPRA (W. Coblenz program manager) and managed through AFOSR (D. Johnston, G. Witt program managers).

This work made use of the MRL Central Facilities supported by the National Science Foundation under Award No. DMR96-32716.

REFERENCES

1. R.A. Burmeister, G.P. Pighini, and P.E. Greene, *Trans. TMS-AIME* **245**, 587 (1969).
2. K.H. Chang, R. Gibala, D.J. Srolovitz, P.K. Bhattacharya, and J.F. Mansfield, *J. Appl. Phys.* **67**, 4093 (1990).
3. E.A. Fitzgerald, Y.H. Xie, D. Monroe, P.J. Silverman, J.M. Kuo, A.R. Kortan, F.A. Thiei, and B.E. Weir, *J. Vac. Sci. Technol. B* **10** , 1807 (1992).
4. G. Springholz, *Appl. Phys. Lett.* **75**, 3099 (1999).
5. G.H. Olsen, *J. Cryst. Growth* **31**, 223 (1975).
6. S. Kishino, M. Ogirima, and K. Kurata, J. Electrochem. Soc. **119**, 618 (1972).
7. L.B. Freund, *MRS Bulletin* **17** (7), 52 (1992).
8. J. W. P. Hsu, E. A. Fitzgerald, Y. H. Xie, P. J. Silverman, and M. J. Cardillo, *Appl. Phys. Lett.* **61**, 1293 (1992).
9. F. Jonsdottir and L. B. Freund, *Mech. Mater.* **20**, 337 (1995).
10. H. Gao, *J. Mech. Phys. Solids* **42**, 741 (1994).
11. S. Y. Shiryaev, F. Jensen, and J. W. Petersen, *Appl. Phys. Lett.* **64**, 3305 (1994).
12. M. A. Lutz, R. M. Feenstra, F. K. LeGoues, P. M. Mooney, and J. O. Chu, *Appl. Phys. Lett.* **66**, 724 (1995).
13. A. Krost, G. Bauer, and J. Woitok, *Optical Characterization of Epitaxial Semiconductor Layers*, eds. G. Bauer and W. Richter (New York: Springer, 1996), p.287.
14. J.D. Eshelby, *Proc. Roy. Soc.* **A241**, 376 (1957).
15. A.E. Romanov and V.I. Vladimirov, in *Dislocations in Solids*, ed. F.R.N. Nabarro (Elsevier, New York, 1992) Vol. 9, p. 191.

Mat. Res. Soc. Symp. Proc. Vol. 673 © 2001 Materials Research Society

Mechanism for the Reduction of Threading Dislocation Densities in $Si_{0.82}Ge_{0.18}$ Films on Silicon on Insulator Substrates

E.M. Rehder, T.S. Kuan[1], and T.F. Kuech
Materials Science Program, University of Wisconsin-Madison
Madison, WI 53706, U.S.A.
[1]University at Albany, State University of New York
New York, Albany, NY 12222, U.S.A.

ABSTRACT

We have made an extensive study of $Si_{0.82}Ge_{0.18}$ film relaxation on silicon on insulator (SOI) substrates having a top Si layer 40, 70, 330nm, and 10μm thick. SiGe films were deposited with a thickness up to 1.2μm in an ultrahigh vacuum chemical vapor deposition system at 630°C. Following growth, films were characterized by X-ray diffraction and a dislocation revealing etch. The same level of relaxation is reached for each thickness of SiGe film independent of the substrate structure. Accompanying the film relaxation is the development of a tetragonal tensile strain in the thin Si layer of the SOI substrates. This strain reached 0.22% for the 1.2μm film on the 40nm SOI and decreases with SOI thickness. The Si thickness of the SOI substrate also effected the threading dislocation density. For 85% relaxed films the density fell from 7×10^6 pits/cm^2 on bulk Si to 10^3pits/cm^2 for the 40, 70, and 330nm SOI substrates. The buried amorphous layer of the SOI substrate alters the dislocation dynamics by allowing dislocation core spreading or dislocation dissociation. The reduced strain field of these dislocations reduces dislocation interactions and the pinning that results. Without the dislocation pinning, the misfit dislocations can extend longer distances yielding a greatly reduced threading dislocation density.

INTRODUCTION

Relaxed SiGe films on Si substrates increase the lattice constant of the material allowing strain engineering of subsequent layers. For example, Si films can then be grown having a tensile strain. This strain modifies the band structure allowing for high carrier mobilities and electron confinement that is not found in the traditional relaxed Si and compressive SiGe structures.

The SiGe film relaxation occurs by the formation of large numbers of dislocations in the material. Threading dislocations, which terminate on the film surface, can remain and disrupt the operation of electronic devices fabricated in this material. A variety of growth methods have been employed to reduce the threading dislocation density. High and low temperature compositional grading, as well as the use of surfactants, has reduced the threading dislocation density (TDD) to the mid $10^4 cm^{-2}$ for films having 20%Ge [1,2,3,4]. Additionally, SOI substrates have been shown to reduce the TDD in relaxed SiGe films by 10^7 [5,6,7]. In this work we have studied the relaxation process by varying the thickness of the SiGe layer and the top Si layer of the SOI, referred to as Si-SOI. X-ray diffraction was used to determine the strain state of the SiGe and Si layers. The film TDD was measured with a dislocation decorating etch.

EXPERIMENT

The bonded SOI substrates possessed a thin Si layer having a thickness of 20, 50, 310nm, and 10µm. Upon these substrates, and reference bulk Si substrates, a uniform composition $Si_{0.83}Ge_{0.17}$ film was deposited by ultra-high vacuum chemical vapor deposition (UHVCVD). The UHVCVD system utilizes wafer strips 7.6mm x 42mm in size. These strips are heated directly by passing current through the sample. This efficient heating provides for a very clean growth environment. Sample cleaning was achieved by a RCA clean and followed by dry oxidation at 1100°C for 10min. The oxidation forms a clean protective oxide allowing the samples to be handled and stored for extended periods of time. The oxidized wafers are then diced into the necessary strips and stored. Before growth, a 5 minute etch in 10%HF removes the oxide and provides a clean sample surface for growth.

The sample is then moved into the load-locked growth system. Flow of SiH_4 (10% in H_2) is initiated, and the sample is annealed at 830°C for 30 seconds to desorb any remaining contaminants. The sample temperature is reduced to 630°C for the remainder of the growth. A clean Si-SiGe interface is obtained by first depositing 20nm of Si, which increases the Si-SOI thickness. SiGe layer growth is carried out with the addition of GeH_4 (5% in H_2). The growth pressure of 23mTorr yields a growth rate of 6.9nm/minute. Films are grown to the desired thickness(340, 765, or 1200nm) and then capped with 4nm of Si to smooth the surface slightly. The Si cap layer does not affect the relaxation process but improves the X-ray diffraction and the interpretation of the etch pit studies. The specific sample will be referenced by the thickness of the thin top Si layer of the SOI substrate after adding the additional 20nm, e.g. 70nm SOI.

Layer strain was determined by X-ray diffraction. Triple crystal reciprocal space maps of the (004) peaks separated the strain, tilt, and mosaic spread components of the films on the SOI substrates. On the 40 and 70nm SOI substrates the low signal intensity of the Si-SOI layer required scans and reciprocal space maps to be taken with a narrow detector slit instead of the analyzer crystal.

The etch pit studies were performed with a mixture of CrO_3-H_2O-HF [8]. The etch contains equal parts of a 1.5Molal CrO_3 solution and 49% HF. The solution etches the SiGe film and Si at 0.6µm/min. An etch time of 45 seconds was used. This proved long enough to provide defect delineation and prevent etching completely through the film. Observations of the etch pits were made with Nomarski phase contrast optical microscopy and scanning electron microscopy.

RESULTS AND DISCUSSION

The relaxation of the SiGe film on the 330nm SOI substrate is shown in Fig. 1. The reciprocal space maps are centered on the Si handle wafer. The Si-SOI layer occurs at nearly the same θ-2θ position, but is tilted slightly due to the wafer bonding process. As the film is grown thicker, relaxation is evident as the SiGe peak moves towards the Si and broadens. The SiGe relaxation with film thickness is common for all of the SOI and the bulk Si substrates (Fig. 2). In addition, the Si-SOI layer develops tensile strain evident in Fig. 1 (c) and (d) as motion of the Si-SOI peak towards positive θ-2θ. This strain develops in the 40, 70, and 330nm SOI samples when the SiGe film has undergone significant relaxation (Fig. 3).

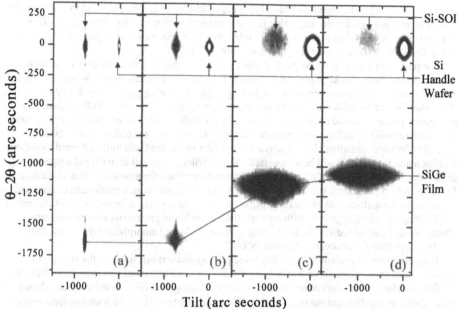

Fig. 1. Triple crystal reciprocal space maps of the (004) peak of the 330nm SOI samples. Four films having different film thickness are shown (a), 150nm; (b), 340nm; (c), 765nm; (d), 1.2μm. The logarithmic intensity scale maximum is adjusted to highlight the SiGe peak, resulting in the truncation of Si handle wafer peak.

While the relaxation of the 1.2μm SiGe film is comparable between the SOI and the bulk Si substrates, the TDD changes dramatically (Fig. 4). The few pits on the 70 and 330nm SOI substrates are circled. The larger pits present in the 70nm SOI image are from growth contamination and are easily separated from the dislocation pits due to their size difference. The etch pit density (EPD) on the 70 and 330nm SOI is below $10^3 cm^{-2}$. The 40nm SOI is not shown but also yielded an EPD below $10^3 cm^{-2}$. The EPD has increased to $1 \times 10^6 cm^{-2}$ on the 10μm SOI substrate, and has reached $7 \times 10^6 cm^{-2}$ for the bulk Si substrate. Similar values were also observed on the samples having a 69% relaxed 765nm thick SiGe film.

The initial explanation for the reduction in TDD of SiGe films on SOI substrates by Legoues et al., involved the SiO_2 layer behaving plastically and slippage occurring at the Si-SiO_2 interface [6]. The mismatch strain would then be partitioned between the Si and the SiGe layers elastically, leading to dislocation formation in the thinner, strained Si layer. In Figure 1, (a) and (b), we do not observe strain transfer to the Si-SOI layer. Also, as the dislocations form beyond the critical thickness in the 765nm and 1.2μm films, strain develops in the Si-SOI as presented in Fig. 3. Fig. 3 also has a plot of the strain expected from elastic strain partitioning, which is more than 3x the observed strain. Therefore we do not see evidence that strain partitioning is occurring.

An alternative mechanism describing this data, specifically the low TDD and the strain in the Si-SOI layer, has been developed, which is also consistent with the data of Refs. 5,6,7. This mechanism is an extension of the SiGe film relaxation by nucleation and propagation of dislocations that are present during growth on bulk Si[9,10].

The pile-up of dislocations on parallel glide planes and reactive blocking of perpendicular dislocations has been found to pin dislocations at high levels of excess strain. When pinning is avoided it occurs with a segment of the impeding dislocation being pushed into the substrate by a dislocation-dislocation interaction with the approaching dislocation. On an SOI substrate, the dislocation pushed into the substrate may reach the amorphous SiO_2 layer. When the dislocation reaches the Si-SiO_2 interface, the associated strain field will decrease either by dissociation [11,12] or by core spreading [13]. The reduction in the strain field will diminish the dislocation repulsion and the pinning will be suppressed. Misfit dislocations will then extend longer distances. The total length of all misfit dislocations determines the amount of film strain relaxed. For a given film relaxation, the reduced pinning results in fewer, longer misfit dislocations and fewer threading dislocations in the film. Also, as the misfit extends, it increases its likelihood of annihilating threading segments with another dislocation having matching burgers vectors. These two dislocation reduction effects, arising from the buried amorphous layer of the SOI substrate, produce the dramatic reduction in TDD.

Image forces will present an attractive force pulling dislocations towards the softer oxide. However, this short-range force will only be significant at a distance of less than 40nm from the Si-SiO_2 interface. The dislocation interactions active during film relaxation have been shown capable of moving dislocations as far as 10μm into the substrate [14]. This strong, long-range interaction is clearly the dominating cause of the dislocations reaching the buried oxide.

The reduced dislocation pinning results in more dislocation interactions and more dislocations extending into the substrate, producing a higher dislocation density in the Si-SOI compared with bulk Si. The strain in the Si-SOI layer that is observed is a direct result of these

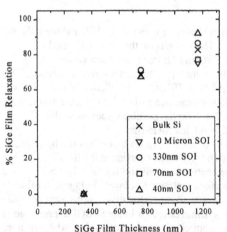

Fig. 2. SiGe film relaxation as the film thickness is increased. The extent of relaxation is independent of substrate.

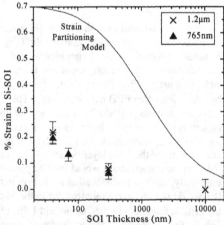

Fig. 3. The dependence of the strain in the Si-SOI layer as a function of SiGe film thickness and Si-SOI thickness. The data indicates that the strain-partitioning model in not applicable to this case.

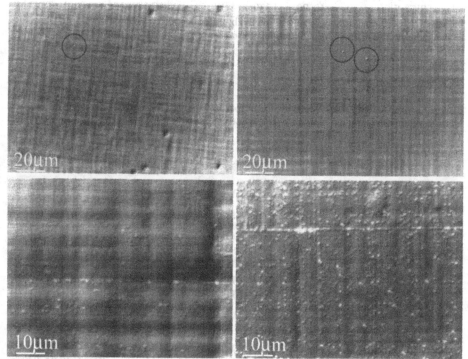

Fig. 4. Influence of the SOI in reducing the threading dislocation density. The rings circle the 3 pits in the top two images. Top left, 70nm SOI; top right, 330nm SOI; bottom left, 10μm SOI; bottom right, bulk Si.

dislocations in the Si-SOI layer. The dislocations produce displacements that increase the in-plane lattice constant, relaxing the SiGe film. The relaxation of the SiGe tetragonal distortion reduces the SiGe out-of-plane lattice constant. This moves the θ-2θ angle of the (004) diffraction spot in the observed positive direction. Conversely, when the dislocations enter the Si-SOI layer the in-plane lattice constant is again increased, which now acts to produce the tensile, tetragonal distortion in the Si-SOI layer. The out-of-plane lattice constant of the Si-SOI layer decreases, accounting for the movement of the diffracted spot toward higher θ-2θ.

CONCLUSION

We have grown a series of $Si_{0.83}Ge_{0.17}$ films up to 1.2μm on SOI substrates whose Si-SOI thickness increases from 40nm to 10μm and on bulk Si substrates. We observe similar amounts of relaxation taking place for a given film thickness, independent of substrate structure. During relaxation, strain develops in the Si-SOI layer. This strain is as high as 0.22% in the 40nm SOI substrate and decreases as the Si-SOI thickness increases. The TDD is less than $10^3 cm^{-2}$ for the relaxed SiGe films grown·on 40, 70, and 330nm SOI substrates. On the 10μm SOI, the TDD has increased to $1 \times 10^6 cm^{-2}$ versus $7 \times 10^6 cm^{-2}$ on the bulk Si substrate. We have developed a model of the dislocation dynamics, which accounts for our observations of a reduced TDD and strain in the Si-SOI layer.

ACKNOWLEDGEMENTS

This work was funded by the NSF-MRSEC at the University of Wisconsin-Madison and the ONR Program on Compliant substrates (C. Wood). In addition, A. Lal generously supplied the 10μm SOI substrates used in this work.

REFERENCES

1. E.A. Fitzgerald, Y.-H. Xie, M.L. Green, D. Brasen, A.R. Kortan, J. Michel, Y.-J Mii, and B.E. Weir, Appl. Phys. Lett. **59**, 811 (1991).
2. P.I. Gaiduk, A.N. Larsen, and J.L. Hansen, Thin Solid Films **367**, 120 (2000).
3. F.K. Legoues, B.S. Meyerson, and J.F. Morar, Phys. Rev. Lett. **66**, 2903 (1991).
4. J.L. Liu, C.D. Moore, G.D. U'Ren, Y.H. Luo, Y. Lu, G. Jin, S.G. Thomas, M.S. Goorsky, and K.L. Wang, Appl. Phys. Lett. **75**, 1586 (1999)
5. A.R. Powell, S.S. Iyer, and F.K. LeGoues, Appl. Phys. Lett. **64**, 1856 (1994).
6. F.K. LeGoues, A. Powell, and S.S. Iyer, J. Appl. Phys. **75**, 7240 (1994).
7. Z. Yang, J. Alperin, W.I. Wang, S.S. Iyer, T.S. Kuan, and F. Semendy, J. Vac. Sci. Technol. B **16** 1489 (1998).
8. K.H. Yang, J. Electrochem. Soc. **131**, 1140 (1984).
9. K.W. Schwarz, J. Appl. Phys. **85**, 120 (1999)
10. E.A. Stach, K.W. Schwarz, R. Hull, F.M. Ross, and R.M. Tromp, Phys. Rev. Lett. **84**, 947 (2000).
11. M. Dollar and H. Gleiter, Scripta Metallurgica, **20**, 275 (1986).
12. J.S. Liu and R.W. Balluffi in Thin Films and Interfaces II, edited by J.E.E. Baglin, D.R. Campbell, W.K. Chu, (Mater. Res. Soc. Proc. **25**, Pittsburgh, PA, 1983) pp 261-272.
13. P. Mullner and E. Arzt, in Thin Films – Stresses and Mechanical Properties VII, edited by R.C. Cammarata, M. Nastasi, E.P. Busso, and W.C. Oliver, (Mater. Res. Soc. Proc. **505**, Warrendale, PA, 1998) pp 149-54.
14. F.K. Legoues, K. Eberl, and S.S. Iyer, Appl. Phys. Lett. **60**, 2862 (1992).

Mat. Res. Soc. Symp. Proc. Vol. 673 © 2001 Materials Research Society

TEM Study of strain states in III-V semiconductor epitaxial layers.

André ROCHER, Anne PONCHET, Stéphanie BLANC* and Chantal FONTAINE*.
Centre d'Elaboration de Matériaux et d'Etudes Structurales, CEMES/CNRS, BP 4347,
Toulouse, F-31055, France.
*Laboratoire d'Analyse et d'Architecture des Systèmes, LAAS/CNRS, 7 av. Colonel Roche,
Toulouse, F-31077, France.

ABSTRACT

The strain states induced by a lattice mismatch in epitaxial systems have been studied by Transmission Electron Microscopy (TEM) using the moiré fringe technique on plane view samples. For the GaSb/(001)GaAs system, moiré patterns suggest that the GaSb layer is free of stress and homogeneously relaxed by a perfect square array of Lomer dislocations. A 10 nm thick layer of GaInAs (20% In concentration) grown on (001)GaAs does not give any moiré fringes for all low-index Bragg reflections: this result indicates that the effective misfit strain does not correspond to the theoretical one described by the elastic theory. Segregation effects are expected to play an important role in the relaxation of the misfit strain.

INTRODUCTION

The strain induced by a lattice mismatch in epitaxial systems is known to control the main physical properties of quantum wells. For epitaxial structures, such strains are assumed to be perfectly pseudomorphic when misfit dislocations are absent at the interface.

A TEM technique for studying the misfit strain is described in this article. It is based on the moiré phenomenon, which is very sensitive for evaluating the lattice parameter of an epitaxial layer relative to that of the substrate. The crystalline quality of a layer is directly characterized, with the conventional TEM resolution, by the homogeneity of the moiré pattern. Comparison of the experimental and theoretical values of the lattice parameters allows us to evaluate locally the epitaxial strain.

Two systems are studied: the GaSb/(001)GaAs, fully relaxed by a perfect misfit dislocation network, which gives rise to well defined moiré patterns, and GaInAs/(001)GaAs, where the strain has to be measured.

ANALYSIS OF MISFIT STRAIN

All models of epitaxial strain are based on elasticity theory, the crystalline perfection of the epitaxial layer and the hypothesis of a perfect interface with perfect bonding. For cubic crystals, the lattice mismatch is defined as $m = (a - b)/a$, where a and b are the film and substrate lattice parameters. At the onset of epitaxy, the film takes the same crystalline orientation as the substrate. When the film is assumed to be pseudomorphic, i.e. elastically strained (Fig.1), the planes normal to the interface have exactly the same spacing in both the layer and the substrate. For the (001) interface, the distortion is tetragonal and the strain $\varepsilon_{//}$ and $\varepsilon_{\perp}$ components calculated by the elasticity theory are given by:

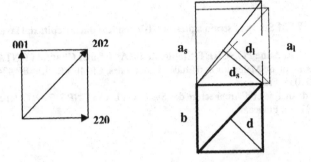

Figure 1: Diagram of the pseudomorphic (solid lines) and relaxed (dotted lines) unit cells (m>0).

$$\varepsilon_{//} = \varepsilon_{100} = \varepsilon_{010} = -m \qquad \text{and} \qquad \varepsilon_{\perp} = \varepsilon_{001} = -\varepsilon_{//} * (2C_{12}/C_{11})$$

In III–V compounds, $\varepsilon_{\perp} \approx 0.9\, m$.

The $\varepsilon_{\perp}$ and $\varepsilon_{//}$ components can be studied by the moiré technique as suggested by Williams and Carter [1]. The moiré effect is given by the interferences resulting from the passage of the electron beam through the substrate and the epitaxial layer. The parallel moiré fringe spacing for an epitaxial system, D_m, is given by the general equation:

$$D_m = d*d_l / (d_l - d) \qquad (1)$$

where d and d_l are the planar distances for the substrate and epilayer. The related contrast gives a very sensitive measure of the planar distances especially for a lattice mismatch of a few percent. For a fully relaxed system, the moiré spacing for a typical {220} reflection, d = 2Å, is 20 nm for a lattice mismatch m of 1% and 2.5 nm for m = 8%.

The pseudomorphic structure, drawn Fig.1, shows that the planes normal to the interface have the same spacing for both substrate and film: the moiré fringe spacing is infinite and there is no moiré pattern for the {220} reflections. An observation using the (004) reflection prevents us from making any comparative analysis of lattice spacings in the substrate and the epilayer. Only the planes inclined with the respect to the interface only are capable of providing information about the strain component normal to the interface. For instance, the pseudomorphic (202) planes are characterized by a distance, d_s (Fig.1), given by in a first order approximation by:

$$d_s = d*(1 + (m + \varepsilon_{\perp})/2). \qquad (2)$$

and the (202) moiré fringe spacing related to the strained structure is given by:

$$D_m \sim d/m$$

this moiré spacing should be equal to about 20 nm for a 1 % lattice mismatch.

In conclusion, a pseudomorphic system should, according the elasticity theory, be characterized by the absence of a moiré pattern for reflections normal to the interface and by a well-defined moiré pattern for inclined reflections such as {202} or {111}. The measured moiré spacing, D_m, related to this inclined reflection allow us to determine the planar distance, d_i, through the relations:

$$d_s = D_m*d / (D_m - d) \qquad \text{and} \qquad \delta d_s / d_s = (d / D_m) * \delta D_m / D_m$$

The accuracy, for m = 1 %, d = 2 Å and $\delta D_m / D_m$ = 5 %, is $\delta d_i / d_i$ = 5*10^-4. This method allows us to obtain accurate values of d_s and to evaluate the applied strain field of the epitaxial film.

The plane view observations are very convenient for studying the crystalline quality of a thin epilayer: the uniformity of the moiré pattern with a well defined contrast indicates that the film is homogeneous. Weak contrast indicates that the film strain does not conform to an ideal elastic description. Measurement of three fringe spacings obtained with two reflections normal and one inclined to the interface is sufficient to determine independently the three basic distances and hence establish the strain state of the epilayer.

EXPERIMENTAL RESULTS

The specimens investigated are epitaxial films thinner than 10 nm allowing plan view observation. TEM samples are prepared by thinning the substrate from the back side without any etching of the epilayer. The observation is mainly performed under dark field in two-beam conditions in order to obtain suitable moiré patterns for which the fringe spacing can be easily measured.

Two systems have been studied: i) the GaSb/(001)GaAs for verifying the capability of the method; ii) the GaInAs/(001)GaAs with 20 % of indium concentration for evaluating the strain state.

GaSb/(001)GaAs

This system with a 8% lattice mismatch has been grown by MBE under the optimum conditions described by Raisin et al [2]. With a nominal thickness of 7 nm, the film appears not to be continuous but consists of isolated islands. Most of the GaSb islands are characterized by clear and simple moiré patterns as observed in Fig.2. The dark line observed in Fig.2a has been characterized as a stacking fault induced by an atomic step at the surface of the substrate [3]. The moiré fringes related to the {200} and {111} reflections, shown in Fig.2, are uniform with well defined fringes. The moiré spacings measured in Fig2a,b are listed Table I. The comparison with the calculated values using bulk lattice parameters indicates that the GaSb film is fully relaxed. The corresponding dislocation network, shown in Fig.2c, is constituted by a square grid of Lomer dislocations which perfectly accommodates the 8% lattice mismatch [4].

Table I: comparison between experimental moiré spacing and the value calculated for bulk materials

Reflection	D (exp)	D (calc.)
D_{200}	39 ± 4 Å	38 Å
D_{111}	46 ± 4 Å	44 Å
D_{dislo}	56 ± 4 Å	54 Å

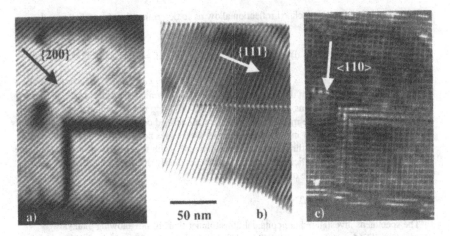

Figure 2: GaSb/(001)GaAs: a) and b) (200) and (111) moiré pattern : note the uniformity of the moiré fringes; c) WB image of the misfit dislocation network, note its regularity.

GaInAs/(001)GaAs

The GaInAs/(001) GaAs systems investigated are grown by solid source MBE at 520°C. An In concentration of 20% leads, following the Vegard law, to a lattice mismatch of about 1.5%. A layer thickness of 10 nm, close to the critical thickness, has been chosen in order to observe the strain field due to the lattice mismatch.

The epilayer thickness, observed by X-TEM, is uniform and equal 10 nm. Figure 3 shows the two plan views taken under the 2 beam conditions in order to evaluate the misfit strain in the GaInAs layer: first, many small features with black/white contrast are seen randomly distributed. For (220) reflection normal to the (001) interface, no moiré fringes and no misfit dislocations are seen in Fig.3a. This result is expected for an elastic strained layer. The dark field image of Fig.3b is obtained with the (202) reflection, inclined to the interface: no moiré pattern is observed. This absence of moiré patterns for all these reflections indicates that the strain state of this epitaxial layer is different from the theoretical one derived from the elastic theory.

DISCUSSION

The different behaviour of these two systems will be discussed in terms of the relaxation and homogeneity of the epilayer.

For GaSb/GaAs, the clear moiré pattern indicates that the GaSb planes are well defined. The perfection of the GaSb lattice planes is due to the island growth with {111} facets: the GaSb grows homoepitaxially on the {111} GaSb facets and dislocations are created at the (001){111} edges, when necessary, by a direct mechanism of creation as discussed by Rocher and Kang [4]. As a result of this direct relaxation, the GaSb film rapidly becomes perfectly crystalline and lattice planes are well enough defined to create the moiré pattern.

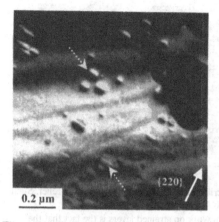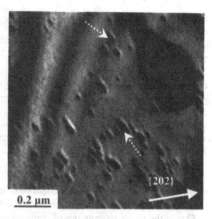

Figure 3: GaInAs/(001)GaAs. (220) and (202) Dark field images. Note the absence of moiré fringes and misfit dislocations. The arrows indicate the same features in the two micrographs.

The behavior of the GaInAs/GaAs system is not so clear. The individual small features, shown Fig.3, have no direct incidence on the elastic misfit stress. The absence of moiré pattern for reflections normal or inclined to the interface indicates that the GaInAs planes are not well defined as would be expected for a homogeneous pseudomorphic layer. Figure.4a is the schematic representation of an ideal epilayer with an average composition of the III atoms, here 25% In concentration, where well defined the lattice planes are able to give rise to a moiré pattern.

A 1.5% lattice mismatch corresponds to an applied biaxial stress of about 1.5 GPa. This value is much higher than the minimal stress needed to induce the movement of dislocations at typical MBE temperatures (> 500°C), as discussed by Yonenaga for bulk GaAs and InP [5]. The first deposited monolayers cannot be considered as homogeneous: STM observations and X-Ray analysis, performed on GaInAs/GaAs by Garreau et al [6], show that the first atomic layer is very heterogeneous with many configurations. The heterogeneity of the first layers can originate from the difference in the chemical nature of the III atoms and from the non-uniformity of the initial surface structure due to its reconstruction and its natural defects such as atomic steps. Under these conditions, it would be most surprising if an epitaxial process could build a homogeneous strained layer from the heterogeneous first monolayers.

The In segregation is known to occur at the start of the growth of GaInAs/GaAs so that the chemical composition of the initial layer is different from its nominal value. Because of this mechanism, the In concentration evolves monotonically from 0 to its nominal value over a perturbed zone of about 3 nm thick near the interface, as estimated using the thermodynamical model of segregation proposed by Moison et al [7]. When its chemical composition becomes nominal, as shown on the top of the Fig.4b, the ternary compound is not homogeneous enough to give rise a well defined lattice planes able to provide moiré pattern. The progressive incorporation of In acts as a direct chemical relaxation. This transition zone between the

Figure 4: schematic representation of a GaInAs layer: a) the ideal strained layer with a homogenous composition of In and Ga atoms; b) the composition gradient induced by In segregation (In atoms, in dark). Far from the interface, the In concentration becomes nominal but not uniform. The free surface where the In concentration is higher than the nominal one, is not represented here.

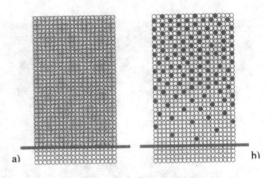

substrate and the uniform epilayer could be described as a chemical buffer for strain relief. Such transition zones have also been observed in GaAs/InP by Rocher and Snoeck [8].

One of the major assumptions of conventional work on strained layers is the fact that the absence of misfit dislocation contrast indicates that the epitaxial film is perfectly strained. Our work suggest that the absence of misfit dislocation image is not a sufficient criterion to characterize a strained structure.

CONCLUSION

This moiré TEM analysis has been applied with some success for measuring lattice parameters of relaxed films. The absence of moiré pattern for GaInAs/GaAs indicates that the epitaxial layer is not strained as described by the classical elastic model of epitaxy. This result can be explained by the presence of a transient zone caused by a chemical composition gradient induced by a segregation effect at the level of the interface. Work is in progress to get a better understanding of the effective role of this transition zone on the strain profile in heteroepitaxy.

REFERENCES

1. D.B. Williams and C.B. Carter, Transmission Electron Microscopy, Imaging **III**, (Plenum,1996), p. 444.
2. C. Raisin, B. Saguintaah, H. Tetegmouse, L. Lassabatère, B. Girault and C. Allibert, Ann. telecommun. **41**, 50 (1986).
3. A. Rocher, Inst. Phys. Conf. Ser. **157**, 153 (1997).
4. A. Rocher, J.M. Kang, Inst. Phys. Conf. Ser. **146**, 135 (1995).
5. I. Yonenaga, J. Phys. III. **7**, 1435 (1997).
6. Y. Garreau, K. Aid, M. Sauvage-Simkin, R. Pinchaux, C.F. Mc Conville, T.S. Jones, J.L. Sudijino and E.S. Tok, Phys. Rev B, **58**, 16177, (1998).
7. B. Moison, C. Guille, F. Houzay, F.Barthe, M Van Rompay, Phys. Rev. **B 40**-9, 6149-6162 (1989).
8. A. Rocher, E. Snoeck, MRS Symp. Poc. Vol **594**, 169 (2000).

Mat. Res. Soc. Symp. Proc. Vol. 673 © 2001 Materials Research Society

A Kinetic Model for the Strain Relaxation in Heteroepitaxial Thin Film Systems

Y.W. Zhang, T.C. Wang[1] and S.J. Chua
Institute of Materials Research and Engineering
National University of Singapore, Singapore, 117602.
[1]LNM, Institute of Mechanics, CAS, P.R. China, 10080.

ABSTRACT

A kinetic model is presented to simulate the strain relaxation in the $Ge_xSi_{1-x}/Si(100)$ systems. In the model, the nucleation, propagation and annihilation of threading dislocations, the interaction between threading dislocations and misfit dislocations, and surface roughness are taken into account. The model reproduces a wide range of experimental results. The implications of its predictions on the threading dislocation reduction during the growth processes of the heteoepitaxial thin film systems are discussed.

INTRODUCTION

The accumulation of experimental data on the heteroepitaxial growth of the Ge_xSi_{1-x}/Si system [1-5] has inspired many researchers to establish models to interpret these data. Initially, energetic models were proposed [6,7], however, further studies indicated that the strain relaxation process was much more complicated than anticipated and these energetic models only revealed the relaxation process correctly at high temperatures. At lower and intermediate temperature ranges, dislocation kinetics must be taken into account [8,9].

Most of previous kinetic models [2,3,5,10] assumed that the strain relaxation was dependent on the velocity of threading dislocations, the dislocations existing in the film and the excess stress for threading dislocation propagation. Our model is basically along the similar line. However, we argue that the excess stress for threading dislocation nucleation is different from that of threading dislocation propagation. Based on this point, an excess stress for threading dislocation nucleation is derived. Moreover, Fitzgerald et al [11] have shown that threading dislocation density is very sensitive to the surface roughness of the film. So in our model, the effect of surface roughness on threading dislocation nucleation is also taken into account.

The present paper is organized in the following manner. First we present our kinetic model for the propagation, nucleation and annihilation of threading dislocations, the hardening effect of misfit dislocations, and the roughness of a film surface. Next we compare our modeling results with a wide range of experimental results. And next the implications of the present model in the reduction of threading dislocation density are discussed. Finally we summarize our main results.

KINETIC MODEL

Consider a Ge_xSi_{1-x} thin film with thickness h, shear elastic modulus μ, and Poisson ratio ν, grown on a semi-infinite large Si substrate. Their mismatch strain is ε_0. For simplicity, we assume that the film and the substrate have same elastic modulus and Poisson ratio. Now we formulate the dislocation kinetic equations.

a. Threading dislocation propagation.

The total stress for threading dislocation propagation is,

$$\sigma_{tp} = \sigma_{exv} - \sigma_r , (1)$$

where, σ_{exv} is the excess stress for the $60°$ dislocation propagation [12], and has the following form,

$$\sigma_{exv} = c\mu(\varepsilon_0 - \varepsilon_p) - \frac{c\mu(1-v/4)b}{4\pi(1+v)} \ln\left(\frac{4h}{b}\right), (2)$$

where $c = 2(1+v)/(1-v)$, b is the magnitude of the Burgers vector, ε_p is the plastic strain. The σ_r is the hardening resistance, which is assumed to be

$$\sigma_r = c\mu\alpha H(\varepsilon_p) = c\mu\alpha\left(\frac{\varepsilon_p}{\varepsilon_m}\right)^\beta\left(1 - \text{Tanh}\left(\frac{\gamma\varepsilon_p}{\varepsilon_m}\right)\right), (3)$$

where, ε_m is the maximum mismatch strain, which is 0.0418 for a pure Ge film on a Si substrate, and α, β, and γ are material constants. The power hardening relation $(\varepsilon_p/\varepsilon_m)^\beta$, commonly used in plasticity theory, is applicable to the cases with a high dislocation density. For the present cases with a lower dislocation density, the correction term $1 - \text{Tanh}(\gamma\varepsilon_p/\varepsilon_m)$ is introduced to consider the situation.

If we assume dislocation propagation is a thermally activated event, then the moving velocity of a threading dislocation, v, can be written as,

$$v = v_0(\sigma_{tp}/\mu)^m \exp(-Q_v/kT), (4)$$

where v_0 and m are material parameters, Q_v is the activation energy, k is the Boltzmann constant, and T is the absolute temperature.

b. Threading dislocation nucleation

Here we assume that dislocations can be nucleated only from the film surface. The driving force can be expressed as,

$$\sigma_{tn} = \sigma_{exn} - \sigma_r , (5)$$

where, the excess stress for the dislocation nucleation can be written as,

$$\sigma_{exn} = \frac{Sc\mu(\varepsilon_0 - \varepsilon_p)}{4} - \frac{\mu b}{4\pi(1-v)h}\left\{\left[(2-v)\ln\left(\frac{2\pi h}{b}\right) - 1.758\right] - \frac{(1-2v)}{4(1-v)}\right\}, (6)$$

and σ_r is dislocation nucleation resistance, which has the same form as Eq. (3); $S = 1 + 4\pi A/\lambda$, which is the surface stress concentration at an undulating surface with the wave amplitude A and wavelength λ [14].

If we take threading dislocation nucleation also as a thermally-activated event, the threading dislocation rate can be written as,

$$\dot{\rho}_{nu} = \zeta_0\left(\frac{\sigma_{exn} - \sigma_r}{\mu}\right)^n \exp\left(-\frac{Q_\rho}{kT}\right), (7)$$

where, ζ_0 and n are materials parameters, Q_ρ is the activation energy for threading dislocation nucleation. Eq.7 reflects the fact that both the mechanical driving force and thermal activation contribute to threading dislocation nucleation.

c. Threading dislocation annihilation

If we assume that threading dislocations are randomly distributed, then the annihilation rate should be proportional to ρ^2. We further assume that the annihilation rate is proportional to their moving velocity and their magnitude of Burgers vector, then the annihilation rates can be written as, $\dot{\rho}_{an} = \chi bv\rho^2$, where, χ is the annihilation parameter, which is related to the detail threading dislocation distribution and the interaction distance between threading dislocations.

d. Net threading dislocation production rate

If we neglect the multiplication of threading dislocation inside the film, then the net threading dislocation production rate is $\dot{\rho} = \dot{\rho}_{nu} - \dot{\rho}_{an}$.

e. Plastic strain rate, dislocation spacing, and average moving distance

The plastic strain rate should be proportional to the velocity, the magnitude of Burgers vector and density of the threading dislocation. For a $60°$ dislocation, only half magnitude of its Burgers vector is used to relax mismatch strain. Besides, there are two perpendicular directions for strain relaxation. Therefore, the plastic strain rate can be expressed by $\dot{\varepsilon}_p = \rho bv/4$. Once we know the total plastic strain ε_p, the misfit dislocation separation p can be calculated, i.e.,

$p = b/2\varepsilon_p$. The average dislocation moving distance L can be evaluated by, $L = \int_0^{t_t} vdt$, where

t_t is the total time used in the growth and annealing process.

COMPARISON WITH EXPERIMENTS

We apply the proposed kinetic model to the $Ge_xSi_{1-x}/Si(100)$ systems. Especially we focus our attention on the systematic experiments carried out by Bean et al [2] and Hull et al [3]. For the fundamental parameters, m, n, Q_v and Q_ρ, we simply take from the experimental measurements [3,5], i.e., $m = 2.0$, $n = 2.5$, $Q_v = 1.1\,eV$, and $Q_\rho = 2.2\,eV$. For the elastic property, we choose: $v = 0.22$, $\mu = 60\,GPa$; for the magnitude of Burgers vector, $b = 0.384\,nm$. We use the rest of the materials parameters to fit against some of the experimental results. The obtained results are: $\alpha = 0.0166$, $\beta = 0.198$, $\gamma = 5.0$, $v_0 = 5 \times 10^{12}\,nms^{-1}$, $\zeta_0 = 6 \times 10^9\,nm^{-2}s^{-1}$ and $\chi = 250$. For the surface roughness, we choose $A/\lambda = 0.072$. Now all parameters are known, we test our model on the $Ge_xSi_{1-x}/Si(100)$ systems.

Hull et al [3] have conducted experiments for growth and growth+annealing of $x = 0.24$ structures by using MBE. The pure growth experiment is at 550°C. The growth+anealing experiment is at 550°C growth + 10min 800°C in situ anneal in the MBE. The growth rate $\dot{G} = 0.3\,nms^{-1}$. They have measured the average dislocation spacing p vs epitaxial layer thickness h. Comparison of the experimental and modeling results is shown in Figure 1. It can be seen that our modeling results agree well with their experimental ones.

Hull et al [3] have also carried out experiments for growth of a 35nm-thick Ge_xSi_{1-x}/Si (100) ($x = 0.25$) structure at different temperatures and subsequently annealing for 4min at each temperature. The growth rate was $0.3\,nms^{-1}$. They have measured the average dislocation separation p and threading dislocation density ρ. Comparison of these experimental and modeling results is given in Figure 2 (a) and (b), respectively. It is evident that the agreements are remarkably good.

Equilibrium analysis suggested that once the thickness of a film exceeded a critical thickness, the final elastic strain of the film should only depend on the thickness of the film. However, it was found that the residual elastic strain measured by Bean et al [2] indeed varied with the initial mismatch strain. Therefore we use our model to predict the variation of the final strain with the initial mismatch strain. Bean et al [2] have conducted the experiments at different

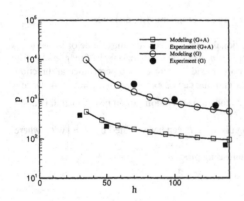

Figure 1 Comparison of average dislocation spacing vs epitaxial film thickness of experiment and modeling for growth 550°C and growth 550°C+10min in situ annealing 800°C of x=0.24 $Ge_xSi_{1-x}/Si(100)$ structure [3]. (G: Growth, A: Annealing).

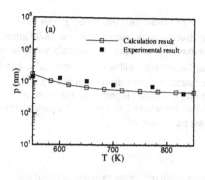

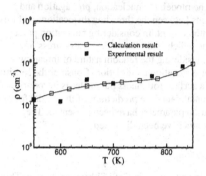

Figure 2 Comparison of experiment and modeling for growth+4min *in situ* annealing at each temperature of x=0.25 Ge$_x$Si$_{1-x}$/Si(100) structure. Experimental data with the filled points was given by given by Hull *et al* [3] and the unfilled points are our modeling results. (a) average dislocation separation, (b) the density of threading dislocation.

thicknesses: 50n, 100nm and 250nm. For each thickness, they used different Ge compositions. Comparison of experimental and modeling results has indicated that although our modeling results of the final elastic strain are a little higher than the corresponding experimental results, however, apparently the present modeling captures the trend the physical phenomenon.

PREDICTIONS OF THE KINETIC MODEL

We have tested our model on the Ge$_x$Si$_{1-x}$/Si systems. Our model has remarkably reproduced a wide range of experimental data. We are confident in using this model to make predictions. We model the experimental scenarios for growth of a 35nm-thick Ge$_x$Si$_{1-x}$/Si (100) ($x = 0.25$) structure and annealing for 4min at each temperature to examine the effects of the surface roughness A/λ, the annihilation parameter χ and the growth rate $\dot{G}$ on the threading dislocation density. Our predictions clearly indicate that if one wants to achieve a threading dislocation density lower than 10^6 cm^{-2}, one has to make the surface roughness parameter A/λ less than 0.04. If together with low temperature growth, the threading dislocation density can be further reduced. We also examine the sensitivity of the annihilation parameter χ. Our model predicts that at the low temperature, the threading dislocation is insensitive to the annihilation parameter χ. With the increase of growth temperature, it is becoming increasingly sensitive. In general, this parameter is much less sensitive than the surface roughness parameter A/λ. For the growth rate, our model predicts that at low temperature, threading dislocation density is insensitive to growth rate. At around 700°C, a lower growth rate is preferred. However, at temperatures higher than 750°C, a higher growth rate is preferred. In general, this parameter is also much less sensitive than the surface roughness parameter A/λ.

SUMMARY

Based on the physical mechanisms of mismatch strain relaxation in heteroepitaxial film growth, a kinetic model has been proposed. In the model, the nucleation, propagation and annihilation of threading dislocations, interactions between the threading dislocations and misfit dislocations, and surface roughness are taken into account. In considering nucleation process, an excess stress for threading dislocation nucleation which is similar to the excess stress for threading dislocation propagation, is derived. Considering the random nature of threading dislocation distribution, we derive a formula for dislocation annihilation. Consider the sensitivity of dislocation nucleation to surface roughness, a surface roughness parameter is introduced. The model has reproduced a wide range of experimental data. The predictions of the model for the surface roughness, growth rate and the annihilation parameter have been presented. The implications and guidelines for growth experiments have been discussed.

REFERENCES

1   R. Hull and E.A. Stach, Strain accommodation and relief in GeSi/Si heteroepitaxy, in Thin film: heteroepitaxial systems, Edited by W.K. Liu and M.B. Santos, World Scientific, Singapore, (1999).
2   R.Hull, J.C. Bean, and C. Buescher, *J. Appl. Phys.* **66**, 5837 (1989).
3   J.C. Bean, L.C. Feldman, A.T. Fiory, S. Nakahara, and I.K. Robinson, *J. Vac. Sci. Technol.* **A2**, 436 (1984).
4   E.A. Fitzgerald, A.Y. Kim, M.T. Currie, T.A. Langdo, G. Taraschi, and M.T. Bulsara, *Mat, Sci. Eng.* **B67**, 53 (1999).
5   D.C. Houghton, *J. Appl. Phys.* **70**, 2136 (1991).
6   J.W. Matthews and A.E. Blakeslee, *J. Cryst. Growth.* **27**, 118 (1974).
7   C.A.B. Ball and J.H. van der Merwe, "The growth of disoclation-free layers", in Dislocations in Solids, F.R.N. Nabarro, Ed. (North-Holland, Amsterdam, 1983), Chapter 27.
8   J.Y. Tsao, Materials fundamentals of molecular bean epitaxy, Boston, Academic Press, (1993).
9   L.B. Freund, *Adv. Appl. Mech.* **30**, 1 (1994).
10   B.W. Dodson and J.Y. Tsao, *Appl. Phys. Lett.* **51**, 1325 (1987).
11   E.A. Fitzgerald, Y.H. Xie, D. Monroe, P.J. Silverman, J.M. Kuo, A.R. Kortan, F.A. Thiel, and B.E. Weir, *J. Vac. Sci. & Technol.* **B10**, 1807 (1992).
12   J.Y. Tsao and B.W. Dodson, *Appl. Phys. Lett.* **53**, 848 (1988).
13   D.J. Bacon and A.G. Crocker, *Philos. Mag.* **12**, 195 (1965).
14   H. Gao, *J. Mech. Phys. Solids* **39**, 443 (1991).

Dislocation Fundamentals:
Observations, Calculations
and Simulations

Mat. Res. Soc. Symp. Proc. Vol. 673 © 2001 Materials Research Society

Dislocation Core Spreading at Interfaces between Crystalline and Amorphous Solids

Huajian Gao[1,2], Lin Zhang[2] and Shefford P. Baker[3]

[1]Max Planck Institute for Metals Research, Seestr. 92, D-70174 Stuttgart, Germany

[2]Department of Mechanical Engineering, Stanford University, Stanford, CA 94305, USA

[3]Department of Materials Science and Engineering, Cornell University, Ithaca, NY 14853, USA

ABSTRACT

A fundamental question addressed here is concerned with the equilibrium structure of a dislocation core at an interface between a crystalline and an amorphous solid. This is motivated by experimental observations that the contrast of dislocations at an interface between a crystalline film and an amorphous substrate disappears under transmission electron microscopy. We have developed a mathematical model to describe the time-dependent behavior of dislocation core spreading as a function of the adhesive strength of the interface. The equilibrium core width and the rate of core spreading are determined in closed form solutions.

INTRODUCTION

Dislocation core structure is known to play an important role on the strength of solids. In thin films deposited on substrates, stresses due to thermal and lattice misfit may be relaxed by dislocation motion through the film. Figure 1 illustrates the motion of a dislocation in a single crystal film subjected to simple biaxial loading. An initial "threading" dislocation-that is, a dislocation running through the thickness of the film-moves on its glide plane in response to the applied stress. If the dislocation is confined to move in the film and not in the substrate, a "misfit" dislocation will be deposited along the film/substrate interface as the threading segment glides on its slip plane.

The concept of misfit dislocations has been used to explain a number of thin film phenomena. Matthews et al. [1] showed that there is a critical film thickness, h_{cr}, below which misfit dislocations cannot form because the energy of the misfit dislocations is greater than the strain energy recovered due to the relaxation associated with their presence. Freund [2] later analyzed the stability of such dislocation structures and derived a rigorous relationship for the critical stress needed to cause dislocation motion in thin films. Based on such analyses, Nix [3] predicted that the yield stress of a film should depend inversely on the film thickness. This prediction has been confirmed experimentally in both semiconductor [4] and metal [5] films and has been used to help explain the fact that very thin films are much stronger than their bulk counterparts.

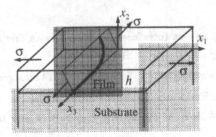

Figure 1: Dislocation motion in a thin film on a substrate leading to the deposition of a misfit dislocation at the interface between the film and the substrate.

The classical concept of dislocation core is based on the model of a line defect in a crystalline structure and does not apply in the case of amorphous solids. Thus a fundamental question is: What is the equilibrium core structure of dislocations at interfaces between crystalline and amorphous solids? At a crystalline/amorphous interface, the core of a dislocation may spread along the interface as illustrated in Figure 2. The substrate has no simple crystallographic relationship to the film and hence may be thought of as a continuum. We have drawn dashed lines on the substrate to mark the original positions of the atomic planes in the film. Depicted in Figure 2 is an edge dislocation with Burgers vector b parallel to the interface climbing towards the substrate under an applied load (Fig 2a) and then spreading its core along the interface (Fig 2b). The dispersed dislocation core may be modeled as an array of infinitesimal dislocations as described by Eshelby [6]. The extent of spreading can be characterized by the half-width, c, of the slipped region.

The simple dislocation core spreading illustrated in Figure 2 requires that the material on one side of the interface be able to slide by small ($< b$) amounts relative to the material on the other side of the interface. This cannot occur over large distances in real coherent interfaces. However, for incoherent interfaces, small relative displacements are possible and dislocation core spreading may occur over large distances. As an illustration consider a crystalline film on an amorphous substrate as shown in Figure 3. Here, the open circles indicate substrate atoms randomly placed in the plane of the paper representing the interface. The filled circles indicate the positions of film atoms in a plane perpendicular to the substrate and some distance from it. We assume that the film atoms bond tightly to the nearest substrate atom. When a shear strain is applied across the interface, we can imagine that some film/substrate bonds are broken, and new, energetically more favorable bonds are formed in their place. Since only a few bonds are relocated at a time, the mean position of the plane can move by an amount that is much less than b.

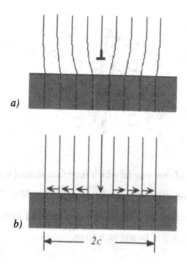

Figure 2: Schematic of edge dislocation climbing towards an interface. The substrate is amorphous. If the adhesion along the interface is perfect, the dislocation core remains compact (a). If it is possible for sliding to occur at the interface, the dislocation core may spread out into the interface (b) to a width $2c$.

A wide range of film/substrate systems have incoherent interfaces and may be susceptible to dislocation core spreading following a mechanism like that illustrated in Figures 2 and 3. Transmission electron microscopy (TEM) investigations have shown dislocations disappearing at the interface between an Al film and its native Al_2O_3 passivation [7] and at the interface between a Cu film and a SiN_2 barrier layer [8]. Here we briefly describe the investigation of Muellner and Arzt[7] who investigated the dislocation structure of Al-2wt%Cu thin films. The thin films were produced by magnetron sputtering onto oxidized silicon substrates. After deposition, the films were annealed at 500°C for 30 minutes and quenched to room temperature. The dislocation structure in metallic films after annealing at high temperature is caused by thermoelastically induced plasticity during cooling. The resulting structure is made up of long, straight, and parallel misfit dislocations along the interface as well as the threading dislocations across the film. In the TEM images, the interface dislocations and threading dislocations are expected to appear simultaneously and with almost the same contrast because they have identical Burgers vector. This is indeed what was observed at the very beginning. However, when the electron beam is kept at a given position for a while, the contrast of the interface dislocations gradually weakens until it finally disappears completely.

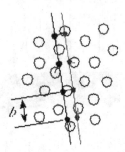

Figure 3: Schematic illustration of one way in which interface sliding may occur. The filled circles represent atoms at the edge of an atomic plane in a crystalline film. The open circles represent bonding sites in an amorphous substrate. The plane can move in increments much less than the Burgers vector by reforming only a few bonds at a time.

After working a few hours on one sample, a wide area does not show any contrast originating from interface dislocations. In Figure 4(a), threading dislocations and interface dislocations are visible. However, the contrast of some of the interface dislocations is already very weak. Figure 4(b) was taken with the very same conditions shortly after the first picture. Almost all contrast of the interface dislocations is gone whereas the contrast of the threading dislocations remains unchanged. The contrast of dislocations is due to the strong localized bending of the lattice planes very near the dislocation core. Thus, the gradual disappearance of the contrast implies that the localized lattice bending gradually diminishes. Muellner and Arzt[7] pointed out that a possible mechanism for this relaxation is dislocation core spreading due to irradiation induced diffusion.

Motivated by the experiments described above, we have developed a mathematical model to describe the time-dependent process of dislocation core spreading along a crystalline/amorphous interface. We consider dislocation spreading along an interface characterized by a shear adhesive strength below which no core spreading occurs, and above which spreading takes place in a viscous manner. We determine the equilibrium core width and the rate of core spreading as a function of interface adhesion. This model is decribed in the following.

MATHEMATICAL MODEL

We consider an edge dislocation spreading along an interface between a crystalline and an amorphous solid, as depicted in Figure 2. For simplicity, possible differences in the elastic constants of the two solids are neglected for the time being. The geometry is defined such that the dislocation

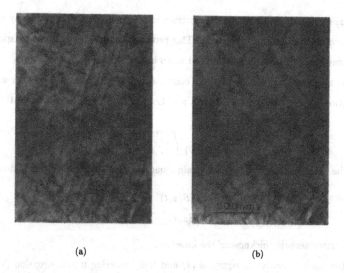

(a)   (b)

Figure 4: TEM pictures of dislocations in an Al 2wt% Cu thin film, taken from Muellner and Arzt [7]. Picture (a) was taken right after the magnification and intensity was adjusted: the contrast of some interface dislocations is visible (straight lines marked with arrows). Picture (b) was taken about one minute after picture (a) with the very same conditions: most of the contrast of interface dislocations is gone.

with Burgers vector b is located at the origin of a rectangular coordinate system (x, y), with interface lying along the $x-$axis. At time $t = 0$, the dislocation reaches the interface and is characterized by a Heavside function

$$S(x,0) = bH(x) = \begin{cases} b & \text{if } x \geq 0, \\ 0 & \text{if } x < 0, \end{cases} \tag{1}$$

for its slip distribution, impling that the dislocation has initially a compact core structure. We assume that deformation near the interface can be described by the following viscous model

$$\dot{\epsilon} = \eta(|\tau| - \tau_0)\text{sign}(\tau), \qquad |\tau| \geq \tau_0, \tag{2}$$

where η is the viscosity depending upon interface properties and temperature, and τ_0 is a threshold stress for slip which will be called the interface adhesive strength.

The stress field around the initial compact core of the dislocation is singular. For a perfectly bonded interface, such stress field will remain unchanged. However, for the "viscoplastic" interface described by (2), the singular stresses will decrease monotonically with time toward the threshold value τ_0. The stress relaxation along the interface is accompanied by slip over any region of the interface where the shear stress exceeds τ_0. This process leads to dislocation core spreading until the shear stress along the interface is reduced to no bigger than τ_0.

We model the spreading dislocation by a continuous dislocation array along the x-axis, with a density function dS/dx. According to Hirth and Lothe[9], the shear stress along the interface is given by

$$\tau(x,t) = \frac{\mu}{2\pi(1-\nu)} \int_{-\infty}^{+\infty} \frac{1}{x-x'} \frac{\partial S(x',t)}{\partial x'} dx' \tag{3}$$

where μ is the shear modulus. The shear strain along the slip plane can be calculated by

$$\epsilon d = \begin{cases} b - S(x,t) & x \geq 0, \\ -S(x,t) & x < 0, \end{cases} \tag{4}$$

where d is a characteristic thickness of the interface.

Taking the time derivative of equation (4) and then inserting it into equation (2) lead to the following equations

$$\begin{cases} -\dfrac{\partial S(x,t)}{\partial t} = \eta d\left[|\tau(x,t)| - \tau_0\right]\mathrm{sign}(\tau) & \text{if } |x| \leq c(t), \\ \dfrac{\partial S(x,t)}{\partial t} = 0 & \text{if } |x| > c(t), \end{cases} \tag{5}$$

where the shear stress $\tau(x,t)$ is given by equation (3), and $c(t)$ is the transient half-width of the dislocation core, corresponding to the x-coordinate of the point at which the shear stress is equal to the critical value τ_0

$$\tau_0 = \tau(c(t),t). \tag{6}$$

Note that the magnitude of the shear stress is less than τ_0 when $x > |c(t)|$.

Substituting the shear stress expression (3) into equations (5) and (6), we obtain the following set of integro-differential governing equations for dislocation core spreading

$$\begin{cases} -\dfrac{\partial S(x,t)}{\partial t} = \eta d\left[\dfrac{\mu}{2\pi(1-\nu)} \displaystyle\int_{-c(t)}^{c(t)} \dfrac{1}{x-x'} \dfrac{\partial S(x',t)}{\partial x'} dx' - \tau_0'\right] \\ \tau_0 = \dfrac{\mu}{2\pi(1-\nu)} \displaystyle\int_{-c(t)}^{c(t)} \dfrac{1}{c(t)-x'} \dfrac{\partial S(x',t)}{\partial x'} dx' \\ S(x,0) = bH(x), \quad S(c(t),t) = b, \quad S(-c(t),t) = 0 \end{cases} \tag{7}$$

where $\tau_0' = \tau_0$ for $0 \leq x \leq c(t)$, and $\tau_0' = -\tau_0$ for $-c(t) \leq x < 0$. Equation (7) is a strongly coupled elasticity and viscous slip problem with dynamically moving boundary conditions. The method adopted to solve such equations will be discussed shortly.

EQUILIBRIUM SOLUTION

We first derive the equilibrium solution to the dispersed dislocation core, in which case equation
(7) is reduced to

$$
\begin{cases}
\displaystyle \int_{-c_e}^{c_e} \frac{1}{x-x'} \frac{\partial S_e(x')}{\partial x'} dx' = \frac{2\pi(1-\nu)\tau_0'}{\mu}, \\
S_e(c_e) = b, \quad S_e(-c_e) = 0,
\end{cases}
\tag{8}
$$

where the subscript 'e' denotes equilibrium state value. Using the method of integral equation
inversion [10], we obtain a closed form solution

$$
\frac{\partial S_e}{\partial x} = \frac{1}{\pi}\left(1 - \frac{4(1-\nu)\tau_0 c_e}{\mu b}\right)\frac{b}{\sqrt{c_e^2 - x^2}} + \frac{4(1-\nu)\tau_0}{\pi\mu}\ln\left|\frac{\sqrt{c_e + x} + \sqrt{c_e - x}}{\sqrt{c_e + x} - \sqrt{c_e - x}}\right|.
\tag{9}
$$

The core half-width c_e is now determined from the condition that the dislocation density dS_e/dx
should tend to zero as $|x| \to c_e$. This means that the value of c_e should be chosen to remove the
mathematical singularities of the solution at both ends $x = \pm c_e$. Removing the singularities yield
the solution for both the half-width of equilibrium core

$$
c_e = \frac{\mu b}{4(1-\nu)\tau_0}
\tag{10}
$$

and the equilibrium dislocation density

$$
\frac{\partial S_e}{\partial x} = \frac{4(1-\nu)\tau_0}{\pi\mu}\ln\left|\frac{\sqrt{c_e + x} + \sqrt{c_e - x}}{\sqrt{c_e + x} - \sqrt{c_e - x}}\right|.
\tag{11}
$$

These solutions show that the final core width $w_e = 2c_e$ is inversely proportional to the interface
adhesion strength τ_0. The stronger the interface, the more compact the dislocation core. After
integration, the expression for the slip function S_e is found to be

$$
S_e = \frac{4(1-\nu)\tau_0}{\pi\mu} x \ln\left|\frac{\sqrt{c_e + x} + \sqrt{c_e - x}}{\sqrt{c_e + x} - \sqrt{c_e - x}}\right| + \frac{b}{\pi}\arcsin\frac{x}{c_e} + \frac{b}{2}.
\tag{12}
$$

TRANSIENT SOLUTION

We have developed an implicit finite difference scheme for a transient solution to equation (7). The
details can be found in [11]. We define the following normalizing parameters

$$
\bar{x} = x/c(t), \quad \bar{c}(t) = c(t)/b, \quad \bar{S} = S/b, \quad \bar{t} = t/t_0, \quad t_0 = \mu\eta d/(2\pi(1-\nu)b),
\tag{13}
$$

and a dimensionless, "relative" strength of the interface

$$
\tau_{rel} = 2\pi(1-\nu)\tau_0/\mu.
\tag{14}
$$

We obtain the results of dislocation core spreading for several different relative interface strengths

$\tau_{rel} = 2\pi(1-\nu)\tau_0/\mu$ using 200 grid points. Only the results for $\tau_{rel} = 0.1$ $((1-\nu)\tau_0/\mu = 1.6 \times 10^{-2})$ will be discussed here. Figure 5 shows the evolution of the dislocation displacement S, the density dS/dx and the shear stress τ within the dislocation core ($|x| \leq c(t)$). The displacement S evolves from the initial Heaviside function to the equilibrium solution (12) while the density function dS/dx decreases quickly from the initial Delta function to the smooth equilibrium solution (11). The singular shear stress in the core area is relaxed to the limiting value τ_0. As time increases, the transient solution tends to the equilibrium solution until the shear stress τ decreases to the critical value τ_0 at which no more slipping occurs in the dislocation core. The figures show that the numerical results at the final stage are in excellent agreement with the exact equilibrium solutions represented by the dashed lines, suggesting our numerical methods are stable and accurate.

Figure 6 shows the evolution of the dislocation half core width $c(t)$. As time increases, $c(t)$ increases monotonically from the initial value $c(0)$ to the equilibrium value c_e. The shape of the curve suggests that $c(t)$ can be fitted into the exponential function

$$c(t) = c_e + [c(0) - c_e]e^{-\frac{t}{a_r t_0}} \tag{15}$$

where a_r is a parameter and $t_0 = \mu\eta d/(2\pi(1-\nu)b)$. The equilibrium value c_e is given by equation (10), and

$$c(0) = \frac{b}{\tau_{rel}} \tag{16}$$

can be readily found by inserting $\partial S(x,0)/\partial x = \delta(x)$ into (6). The characteristic time $a_r t_0$ appearing in (15) is the relaxation time which the compact dislocation core needs to spread out and reach the equilibrium state. For different relative interface strengths we determine the values of the parameter a_r shown in Figure 7. By least squares method, a_r is found to be inversely proportional to the relative interface strength τ_{rel}

$$a_r = \frac{0.1108}{\tau_{rel}} \tag{17}$$

which is represented by the solid line in Figure 7. Substituting (17), (16) and (10) into equation (15), we find that the evolution of $c(t)$ is approximated by

$$c(t) = \frac{\mu b}{4(1-\nu)\tau_0}\left[1 - 0.3634\exp\left(-\frac{356.3(1-\nu)^2\tau_0 b}{\mu^2\eta d}t\right)\right] \tag{18}$$

with the relaxation time $\mu^2\eta d/(356.3(1-\nu)^2\tau_0 b)$.

For very weak interface, equation (18) can be linearized as

$$c(t) = \frac{\mu b}{4(1-\nu)\tau_0}\left(0.6366 + \frac{129.5(1-\nu)^2\tau_0 b}{\mu^2\eta d}t\right). \tag{19}$$

This result shows at a very weak interface the dislocation core expands with a constant velocity.

SUMMARY

We have described a mathematical model of dislocation core spreading at an interface between a crystalline and an amorphous solid. The primary motivation for the study is a need to under dislocation core structure in thin film systems which often involve a crystalline film bonded to an amorphous substrate. We have derived equilibrium solution to the dispersed dislocation core in closed form. We have developed an efficient numerical scheme for calculating the transient behavior of an expanding dislocation core. The main results can be briefly summarized: (1) The equilibrium core width scales inversely with the interface adhesion. The weaker the interface adhesion, the wider the core; (2) For strong interfaces, the core approaches equilibrium width exponentially with time. The transient solution is given explicitly in equation (18); (3) For very weak interfaces, the core expands with constant velocity given in equation (19).

Many thin films systems involve a crystalline films attached to an amorphous substrate or layer such as the native oxide of the film. The phenomenon of dislocation core spreading is expected to occur frequently, especially under relatively high temperatures. For example, as threading dislocations propagate to lay down misfit dislocations in a thin film, the component of Burgers vector parallel to the interface may be subject to dynamic viscous core spreading and, depending upon the interface strength, may significantly affect the thermomechanical behaviors of the film. Controlled experiments are underway to verify such conjectures, and the mathematical modeling developed in this paper will be used to interpret the experimental results in the near future.

References

[1] Matthews, J. W., Mader, S. and Light, T.B. (1970) *J. of Applied Physics*, **41**, 3800-3804.

[2] Freund, L. B. (1987) *Journal of Applied Mech.*, **54**, 553-557.

[3] Nix, W. D. (1989) Mechanical properties of thin films, *Metall. Trans.*, **20A**, 2217-2245.

[4] Bean, J. C., Feldman, L. C., Fiory, A. T., Nakahara, S. and Robinson, I. K. (1984) *J. Vac. Sci. Technol. A*, **2**, 436-440.

[5] Venkatraman, R. and Bravman, J.C. (1992) *J. Mater. Res.*, **7**, 2040.

[6] Eshelby, J.D. (1949) *Phil. Mag.*, **40**, 903.

[7] Muellner, P. and Arzt, E. (1998) in "Thin Films: Stresses and Mechanical Properties VII" ed: R.C. Cammarata, M. Nastasi, E.P. Busso, W.C. Oliver, *Mat. Res. Soc. Symp. Proc.*, **505**, 149-154.

[8] Dehm, G. and Arzt, E. (2000) *Appl. Phys. Lett.*, **77**, 1126-1128.

[9] Hirth, J. P. and Lothe, J. (1982) *Theory of Dislocations*, 2nd edn., John Wiley and Sons.

[10] Muskhelishvili, N. I. (1977) *Singular Integral Equations*, Noordhoff, Leyden.

[11] Zhang, L. (2000) Ph.D. Dissertation, Stanford University, Stanford, CA.

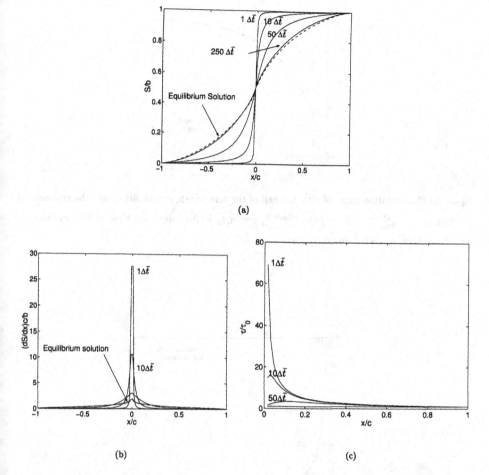

Figure 5: For $\tau_{rel} = 0.1$, the evolution of (a) the displacement S; (b) the density function dS/dx; (c) the shear stress τ within the dislocation core. Here $\bar{t} = t/t_0$, $t_0 = \mu\eta d/(2\pi b)$. The time step $\Delta\bar{t} = 0.01$. Both S and dS/dx tend to the exact solutions of the equilibrium state represented by the dashed lines.

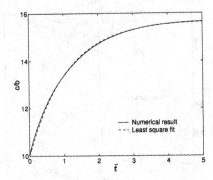

Figure 6: The evolution curve of $c(t)$, the half of the core width, can be fitted into the exponential function $c(t) = \frac{\pi b}{2\tau_{rel}}[1 - (1 - 2/\pi)e^{-t/(a_r t_0)}]$, and $a_r t_0$ is the relaxation time of dislocation core spreading. Here $\tau_{rel} = 0.1$ and $a_r = 1.11$.

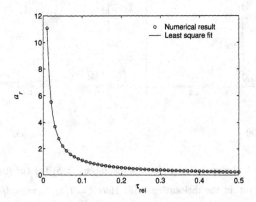

Figure 7: The relaxation time $a_r t_0$ is dependent upon the relative interface strength τ_{rel} through the least square approximation $a_r = 0.1108/\tau_{rel}$. For weaker interface with smaller strength τ_0, the dislocation needs longer time to spread out to the equilibrium state.

Mat. Res. Soc. Symp. Proc. Vol. 673 © 2001 Materials Research Society

Dislocation Networks Strain Fields Induced By Si Wafer Bonding.

J. Eymery, F. Fournel*, K. Rousseau, D. Buttard, F. Leroy, F. Rieutord, J.L. Rouvière.
CEA/Grenoble, Département de Recherche Fondamentale and *LETI/Département des
Technologies Silicium, 17 rue des martyrs, 38054 Grenoble Cedex 9, France.

ABSTRACT

Buried dislocation superlattices are obtained by bonding ultra-thin single crystal Si (001)
films on Si (001) wafers. The twist of two Si wafers induces a regular square grid of dissociated
screw dislocations and the tilt a 1-D array of mixed dislocation. The Burgers vector is a/2 <110>
for both types of dislocation. The atomic displacements and deformations of pure screw and edge
dislocations are calculated with an isotropic elasticity approximation taking into account the free
surface and the thickness of the upper crystal. It is shown by these calculations that the elastic
strain field propagates up to the surface, and quantitative arguments are given to choose the
network period / film thickness ratio.

INTRODUCTION

The self-assembly of semiconductor quantum dots has attracted a lot of interest in recent
years because of their potential applications in novel electronic and optoelectronic devices.
Several experiments have been performed to increase the lateral ordering to control the individual
properties of the dots. One of them consists to achieve a regular nanometric patterning of the
substrate by using a periodic strain field induced by a misfit dislocation network. The control of
the island nucleation by buried dislocations has been demonstrated [1], but the network regularity
was not enough to induce a lateral ordering. An assembling technique using 4-inch (001) direct
wafer bonding and Silicon on Insulator (SOI) wafer [2] has been used recently to obtain high
quality ultra-thin Si crystalline layers on Si substrates. This method is compatible with
microelectronics technology, and allows an accurate control of the network period. Moreover, the
dislocation cores are buried at the interface so that they don't interact directly with the islands. It
has been shown from continuum elasticity calculations [3,4] that the periodic dislocation strain
field close to the surface may be used to obtain a controlled surface patterning and to grow self-
organized quantum dots with a narrow size distribution. The propagation of the strain field to the
surface has been confirmed experimentally by scanning tunneling microscopy. We have observed
on bonded substrate [5] a very regular nanoscale surface patterning and the achievement of the
long range ordering of Si quantum dots on very thin oxide layers.

For a (001) bonded interface, the twist ψ of two Si wafers with flat and parallel (001) surfaces
(i.e. no miscut) induces a regular square grid of pure or dissociated screw dislocations, called
hereafter twist interfacial dislocations (TWIDs). Whereas a tilt angle θ (miscut angle of the
bonded samples) induces 60° or mixed dislocations [2] called tilt interfacial dislocations (TIDs).
The Burgers vector is a/2 <110> for both types of dislocations that are localized at the bonded
interface without emerging to the surface. The interfacial structure has been studied by
Transmission Electron Microscopy (TEM) [6] to determine the nature of the dislocations and
their interactions. X-Ray Specular Reflectivity [7] has been used to study the surface roughness,
the thickness fluctuations of the bonded layer along the full wafer, and the interface quality in

terms of electronic density. Finally, the measurement of the long-range order of the lateral displacement field has been performed by Grazing Incidence X-Ray Diffraction (GIXRD) [8]. In this communication, the general features of the dislocation networks will be discussed, and the maximum surface deformation will be discussed for the pure screw and edge dislocation networks.

EXPERIMENTAL DETAILS

A hydrogen-passivated hydrophobic process is used to bond a standard 4-in. (001) silicon-on-insulator (SOI) on a Si (001) wafer at high temperature (>1100 K) [2]. The advantage of this method is that the thinning of the Si bonded layer is started on a material with a known and homogeneous thickness. The combination of chemical and mechanical thinning gives a 5-200 nm film thickness with a good homogeneity [2,7], much better than what is achieved directly with thick single crystal thinning [8]. The samples are observed by plane-view and cross-section high resolution TEM to check the lack of extended defects and by classical two-beam and weak-beam conditions to visualize the interface dislocations and their regularity [6]. As shown in the plane-view of figure 1 a), two networks are observed. The first one consists of two orthogonal arrays of TWIDs which period is related to the twist angle by the Frank's relation
$$\Lambda^{TWID}=|\vec{b}|/(2\sin(\psi/2))=a_{Si}/[2\sqrt{2}\sin(\psi/2)].$$ Cross sections show that they are dissociated in two 30° partials $(1/2[110]\rightarrow 1/6[211]+1/6[12\bar{1}])$ separated by an intrinsic stacking fault located on a $\{111\}$ plane. The second set is a single array of TIDs lines corresponding to 60° or mixed dislocations. These lines are straight contrary to the meandering observed for lower temperature bonding [9], and the period between the lines is given by $\Lambda^{TWID} = a_{Si}/[2tg(\theta)]$. Note that we observe neither oxide precipitates nor threading dislocation in this sample so that the surface is only perturbed by the dislocation strain.

Figure 1. a) Left: TEM image (weak beam condition g=004) of a bonded sample ($\psi\approx2.75\pm0.07°$, $\theta\approx0.30\pm.06°$). b) Right: Fourier Transform of image a).

Figure 1 b) shows the Fourier Transform of image a). This pattern is composed of two sets of peaks, the first one is coming from the two orthogonal TWID networks, and the second one corresponds to satellites centered on each peaks of the first set (TID network). These peaks, measured by GIXRD [8], are directly related to the periodicity of the strain field. The peak position in reciprocal space can be explained in terms of density perturbation of continuous medium [10]. The scattered amplitude is expanded as the function of the unperturbed density as:

$$F(\mathbf{k}) = F_0(\mathbf{k}) + \frac{1}{(2\pi)^3} F_0(\mathbf{k}) \otimes A(\mathbf{k}) \tag{1}$$

where

$$\rho(\mathbf{r}) = \rho_0(\mathbf{r})[1 + a(\mathbf{r})], \tag{2}$$

$$F(\mathbf{k}) = \int \rho(\mathbf{r}) \exp(-i\mathbf{k} \bullet \mathbf{r}) d\mathbf{r}, \quad F_0(\mathbf{k}) = \int \rho_0(\mathbf{r}) \exp(-i\mathbf{k} \bullet \mathbf{r}) d\mathbf{r}, \quad A(\mathbf{k}) = \int a(\mathbf{r}) \exp(-i\mathbf{k} \bullet \mathbf{r}) d\mathbf{r} \tag{3}$$

With a harmonic plane wave $a(\mathbf{r}) = a_0 \cos[\mathbf{q} \bullet \mathbf{r} - \varphi_a]$, the lattice structure peaks $\mathbf{k}_i$ of $F_0(\mathbf{k})$ will produce extra peaks at $\mathbf{k}_i \pm \mathbf{q}$. The same formalism is applied to the combination of two harmonic perturbations $a(\mathbf{r})$ and $b(\mathbf{r})$ corresponding to the tilt and to the twist of the two crystals:

$$F(\mathbf{k}) = F_0(\mathbf{k}) + \frac{1}{(2\pi)^3} F_0(\mathbf{k}) \otimes [A(\mathbf{k}) + B(\mathbf{k})] + \frac{1}{(2\pi)^6} F_0(\mathbf{k}) \otimes A(\mathbf{k}) \otimes B(\mathbf{k}) \tag{4}$$

The third term shows that the extra peaks $\mathbf{k}_i \pm \mathbf{q}$ of the first modulation will also have satellites at $\mathbf{k}_i \pm \mathbf{q} \pm \mathbf{p}$. An estimation of the satellite intensities can be reached with the calculated displacement fields obtained from the analytical model of Bonnet and Verger-Gaugry [11], but these calculations are out of the scope of this paper. We will present here the displacement and deformation features of the two generic Si wafer-bonding dislocations *i.e.* pure screw and edge cases as the function of the lateral period for a given thin-film thickness. The TIDs are a mixture of these two components. The case of the dissociation of the screw dislocation, playing an important role for the very small crystal thickness and ψ value [6], will not be discussed in this paper.

DISCUSSION

Screw dislocation network

Consider a screw dislocation network of period Λ along x_1 buried at a depth h under the surface. The dislocation line and the Burgers vector is along x_3, x_2 is orthogonal to the surface, and the origin of the axis is chosen at a dislocation core position. The displacement along x_3 of the upper (+) and lower (-) crystals is [11] (this is a real term) :

$$u_3^\pm = i \frac{b}{4\pi} Log \left[\frac{\left(1 - e^{\omega(-2h - ix_1 + x_2)}\right)\left(1 - e^{\pm\omega(-ix_1 - x_2)}\right)}{\left(1 - e^{\omega(-2h + ix_1 + x_2)}\right)\left(1 - e^{\pm\omega(ix_1 - x_2)}\right)} \right] \quad \text{where } \omega = 2\pi/\Lambda \tag{5}$$

The second displacement field $u_1^\pm$, necessary to obtain the in-plane misorientation ψ, is obtained by changing x_1 by $-x_3$ in Eq. 5. For a pure screw dislocation network, the strain field is a shear and $u_2^\pm = 0$. An example of the displacements and of the strain field $\varepsilon_{13}^\pm = (\partial u_1^\pm / \partial x_3 + \partial u_3^\pm / \partial x_1)/2$ calculated at the surface is shown in figure 2. The strain field has one region of maximum dilatation and compression per calculation cell. Due to symmetry consideration, they are located

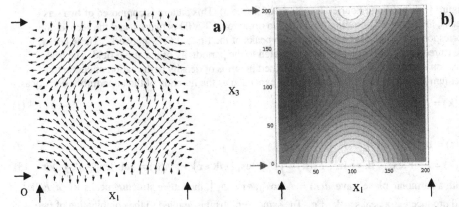

Figure 2. Plane-views of the a) displacement field $u_{1,3}$ calculated at the surface of a Si twist-bonded layer (Λ=200 Å) of thickness 100 Å. The length is proportional to the vector magnitude. b) ε_{13} strain field calculated for the same system. White (black) is a maximum (minimum) value of $\varepsilon_{max} = \pm 1.65 \ 10^{-3}$ for b=a/2 [110]=3.84 Å. Arrows indicate dislocation lines.

to the half of the x_1 and x_2 axis. The magnitude ε_{13}^{max} of the maximum of deformation is studied as the function of Λ for a fixed value of h. Figure 3 shows an example for h=100 Å. ε_{13}^{max} is very low at small Λ and goes to about 6.1 10^{-3} at large Λ. Note that most of the deformation has been obtained for Λ/h=5. The general shape of this curve is due the screening of the dislocation deformations at small Λ. The limit at large Λ tends asymptotically to b / (2 π h).

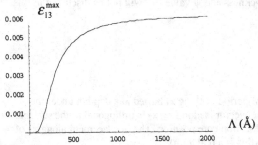

Figure 3. Evolution of the maximum of deformation ε_{13}^{max} (see the white zone in figure 2) for b=a/2 [110]=3.84 Å and h=100 Å as the function of Λ.

Edge dislocation network

For the tilt dislocations, the Burgers vector is along x_2, so that $u_3^{\pm} = 0$. The equations giving $u_1^{\pm}$ and $u_2^{\pm}$ are too long to be given in this paper, but they are illustrated on figure 4 for Λ/h=2.

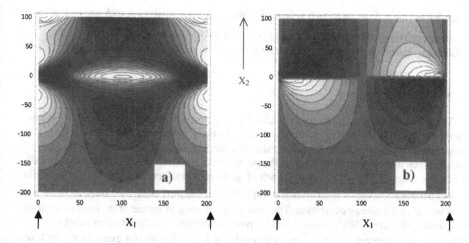

Figure 4. Cross-sections of the a) displacement field $u_1^{\pm}$ and b) $u_2^{\pm}$ of a Si twist-bonded layer (Λ=200 Å) of thickness 100 Å. The scale is linear in the gray level white (black) is a positive (negative) value. The axis origin is at a dislocation core position.

We have to consider the three ε_{11}, ε_{22}, ε_{12} deformations at the surface. As shown in figure 5 for Λ/h=2 and $b=b_2$=1 Å, these three components have the same order of magnitude that is larger than those of the screw dislocation case. But the evolution of the maximum at the function of Λ for h fixed at 100 Å is quite different.

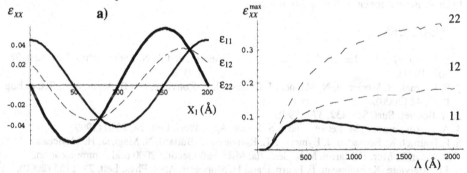

Figure 5. a) Plot of the deformation ε_{xx} (XX=11,22,12) as the function of x_1 (dislocations are at x_1=0 and x_1=200 Å), b) maximum of deformation ε_{xx}^{max} (the x_1 position of the extrema of a) depend on Λ) as the function of Λ for h=100 Å and b_2=1 Å ($b_1=b_3$=0 Å). The axis origin is at a dislocation core position.

As shown in figure 5, ε_{11} goes to 0 at large Λ with a maximum at about $\Lambda/h=5$ whereas ε_{22} and ε_{12} are monotonic and increase to 1/2 and 1/4 respectively. At $\Lambda/h=5$, less than half of the deformation has taken place, but the magnitude is quite large.

CONCLUSIONS

The use of these calculations for the wafer-bonding problem must be done carefully. In real situations, the TIDs are mixed, and the dissociation of the screw dislocation gives an edge component that plays a role for a small h value and small ψ, so that both screw and edge strain must be taken into account. Moreover, if the physical process under study is thermodynamically driven (for example high temperature growth of quantum dots), then the energy of the system is the key parameter to define its evolution (morphology) [3, 4], and all the deformation terms ε_{ij} are involved in the energy expression. This case gives a strong influence of the edge dislocation compared to the screw. But if the process is driven by kinetics, then the situation may be different. Only some components of the deformation will be effective (for example ε_{11} for low temperature epitaxy or chemical etching). These considerations probably explain why dissociated screw dislocation networks are strong enough to drive kinetics mechanisms like low temperature growth or surface patterning [5].

ACKNOWLEDGEMENTS

The authors would like to thank the French "région Rhône-Alpes" to support this research, and the CRG-IF beamline and ESRF to perform X-ray diffraction. The authors are very grateful to Dr R. Bonnet for enlightening discussions.

REFERENCES

1. S. Yu. Shiryaev, F. Jensen, J. L. Hansen, J. W. Petersen, and A. N. Larsen, Phys. Rev. Lett. **78**, 503 (1997).
2. F. Fournel, H. Moriceau, N. Magnéa, J. Eymery, J. L. Rouvière, and B. Aspar, Mat. Sci. & Eng B **73**, 42 (2000).
3. A. Bourret, Surf. Sci. **432**, 37 (1999).
4. A. E. Romanov, P. M. Petroff, and J. S. Speck, Appl. Phys. Lett. **74**, 2280 (1999).
5. F. Fournel, K. Rousseau, J. Eymery, J.L. Rouvière, D. Buttard, N. Magnea, H. Moriceau, B.Aspar, P. Mur, F. Martin, M.N. Semeria, MRS Fall meeting 2000 oral communication.
6. J.L. Rouvière, K. Rousseau, F. Fournel, and H. Moriceau, Appl. Phys. Lett. **77**, 1135 (2000).
7. J. Eymery, F. Fournel, H. Moriceau, F. Rieutord, D. Buttard, and B. Aspar, Appl. Phys. Lett. **75**, 3509 (1999).
8. F. Fournel, H. Moriceau, N. Magnea, J. Eymery, D. Buttard, J. L. Rouvière, K. Rousseau, and B. Aspar, Thin Solid Films **380**, 10 (2000).
9. Q. Y. Tong and U. Gösele, Semiconductor Wafer Bonding (Wiley, New York, 1999).
10. D. C. Champeney, *Fourier Transforms And Their Physical Applications* (Academic Press, London and New York 1973).
11. R. Bonnet and J. L. Verger-Gaugry, Phil. Mag. A **66**, 849 (1992).

Dislocations and Deformation Mechanisms in Thin Metal Films and Multilayers II

Mat. Res. Soc. Symp. Proc. Vol. 673 © 2001 Materials Research Society

Misfit Dislocations in Epitaxial Ni/Cu Bilayer and Cu/Ni/Cu Trilayer Thin Films

Tadashi Yamamoto, Amit Misra, Richard G. Hoagland, Mike Nastasi, Harriet Kung, and John P. Hirth
MST-8, Materials Science and Technology Division
Los Alamos National Laboratory, Los Alamos, NM

ABSTRACT

Misfit dislocations at the interfaces of bilayer (Ni/Cu) and trilayer (Cu/Ni/Cu) thin films were examined by plan-view transmission electron microscopy (TEM). In the bilayers, the spacing of misfit dislocations was measured as a function of Ni layer thickness. The critical thickness, at which misfit dislocations start to appear with the loss of coherency, was found to be between 2 and 5 nm. The spacing of the misfit dislocations decreased with increasing Ni layer thickness and reached a plateau at the thickness of 30 nm. The minimum spacing is observed to be about 20 nm. A g·b analysis of the cross-grid of misfit dislocations revealed 90° Lomer dislocations of the <110>{001} type lying in the (001) interface plane at a relatively large thickness of the Ni layer, but 60° glide dislocations of the <110>{111} type at a relatively small thickness of the Ni layer. In the trilayers, misfit dislocations formed at both interfaces. The spacing of the misfit dislocation is in agreement with that of the bilayers with a similar Ni layer thickness. The misfit dislocation arrays at the two interfaces, having the same line directions, are 60° dislocations with edge components with opposite signs but are displaced with respect to each other in the two different interface planes. This suggests that interactions of the strain fields of the dislocations have a strong influence on their positions at the interface.

INTRODUCTION

Misfit dislocations are known to form at the interface to relieve the elastic strains due to lattice mismatch in heteroepitaxial films [1-7]. In the early stages of the overgrowth, the misfit is entirely accommodated by coherency strains, where two lattice planes are strained in one-to-one registry. As the overgrowth thickness increases, the misfit dislocations start forming at the critical thickness, where the introduction of misfit dislocations lowers the total energy of the system arising from elastic strain and dislocations [1, 2]. Analyses of critical thickness and strain field have been the major subject of both theoretical and experimental studies, where most detailed analyses have been done for semiconductor bilayer couples [2-7]. Once the dislocations form as the thickness exceeds the critical thickness, the dislocation spacing, or strain complementary to the spacing, decreases with either increasing thickness or increasing lattice mismatch [4, 5]. As a function of character, <110>{111} type 60° dislocations are generally observed at relatively small thickness or small lattice mismatch. As the thickness or lattice mismatch increases, the fraction of 60° dislocations decreases and 90° dislocations of the <110>{001} type predominantly form [4, 5]. The 90° dislocation is thought to form by the reaction of two 60° dislocations at the interfaces [6, 8]. For the metal bilayer systems, fcc metals have been studied primarily in the limited thickness range of 1 to 10 nm [9-15]. Coherency strain and misfit dislocation density in the metal bilayers behave with layer thickness in a similar manner to those in the semiconductor bilayers [10, 11]. With respect to character, 60° dislocations have been observed [9, 10], but 90° dislocations have been also reported [11]. No consensus has been reached on this point or other details of misfit dislocation behavior in metal bilayers. The deformation behavior of the multilayer films is also influenced by misfit dislocations through their interactions with gliding dislocations. For example, the misfit dislocations locked in the interfaces can block the motion of gliding dislocations [12]. The spacing between two misfit dislocations at the interfaces determines the stress to move a pinned dislocation by an Orowan type relation. Thus far, there are only limited measurements of spacing vs. layer thickness in the literature. Hence, a comprehensive understanding of the nature of misfit dislocations as a function of layer thickness is of interest fundamentally as well as providing a guide to the fabrication of high-strength multilayers. In this research, we investigate the critical thickness, spacing, configurations, Burgers vectors, and characters of misfit dislocations between two layers by changing the thickness of the overgrowth Ni layer in both Ni/Cu bilayers and a Cu/Ni/Cu trilayer.

EXPERIMENT

Cu and Ni layers were prepared by sequential evaporation of high-pure metals (99.999%) in an electron beam evaporator (ESV 6/UHV, Leybold-Heraeus) and grown cube-on-cube onto the (001) surface of a NaCl crystal, cleaved in air. After the chamber was evacuated to a high vacuum of 7×10^{-8} Torr, the substrate was heated to 350 °C and was held at this temperature for about two hours, which helps to remove the air trapped on the NaCl surface [13]. A 100 nm thick Cu layer was then grown along [001] on the NaCl substrate at 350 °C. After the substrate was cooled to 50 °C, a Ni layer was deposited onto the surface of Cu, again along [001], at the desired thickness from 2 to 70 nm. During the evaporation, the pressure in the chamber rose to between 3×10^{-7} and 3×10^{-6} Torr, but after the deposition, the pressure quickly recovered to the level of the base pressure. The bilayer crystals were removed from the vacuum chamber after they had cooled down to the room temperature. Both Cu and Ni layers were grown at the relatively low growth rate of 0.08 nm/s to minimize the residual stress possibly induced by the bombardment of energetic vapor atoms [14]. Thickness and evaporation rate were monitored by a quartz crystal oscillator (XTC, Inficon). The thickness was also checked by cross-sectional TEM observation. The 70 nm Cu/10 nm Ni/70 nm Cu trilayer was fabricated in a similar manner except that the additional Cu layer was grown at 50 °C right after the deposition of the Ni layer. The films were floated off the NaCl in distilled water at room temperature. The films were mounted onto a 3 mm double-folded Cu grid for the TEM analysis of the misfit dislocations. Both bright field and dark field images were taken on a Philips CM30 microscope operated at 300 kV. A liquid nitrogen cooled anti-contaminator was used throughout the observation.

DISCUSSION

Bilayers

Electron micrographs of 10 nm Ni/100 nm Cu and 70 nm Ni/ 100 nm Cu bilayers, taken with 200 type reflections, are shown in Figure 1. The foils are oriented approximately along the [001] direction. The misfit dislocation lines are straight and parallel to [220] and [2$\overline{2}$0] directions and form cross-grid arrays. The misfit dislocation spacing in the 10 nm thick Ni varies between 30 and 60 nm (Figure 1(A)) and the spacing in 70 nm thick Ni is between 10 and 25 nm (Figure 1(B)). The average spacings in the 10 nm nickel and 70 nm nickel are about 40 and 20 nm respectively. The variation of the spacing is larger in the 10 nm Ni layer than in the 70 nm layer. The average misfit dislocation spacing, determined from several areas for each sample, was plotted as a function of Ni layer thickness (Figure 2). We still observed misfit dislocations for a 5 nm thick Ni layer, but not for a 2 nm nickel layer. This result indicates that the critical thickness is between 2 and 5 nm. As the Ni layer thickness increases, the spacing of misfit dislocations decreases quickly until the thickness reaches 30 nm. As the thickness increases further, the spacing decreases less rapidly and approaches asymptotically to the expected equilibrium value of 20 nm. A **g·b** analysis has been performed to reveal the Burgers vectors and characters of the misfit dislocations. Most of the dislocations observed in the 10 nm Ni/100 nm Cu bilayer sample are 60° glide dislocations of the <110>{111} type. However, in the 30, 50, 70 nm Ni/ 100 nm bilayers we observed that most dislocations are Lomer-Cottrell dislocations of the <110>{001} type lying on the (001) interface plane.

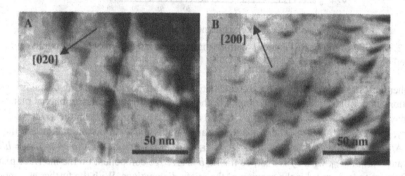

Figure 1. Misfit dislocations at (A) 10 nm Ni and (B) 70 nm Ni epitaxially deposited on 100 nm Cu layer.

The general trend of decreasing misfit dislocation spacing with increasing cap layer thickness is in agreement with earlier observations [3, 4, 10]. Matthews and Crawford [10] studied misfit dislocations in Ni/Cu bilayers prepared by evaporation and measured the elastic strains as a function of Ni thicknesses h from 1 to 9 nm. They observed that the dislocation density increased with increasing thickness of the Ni overgrowth layer, although no measurement of spacing or density of misfit dislocations was conducted. They also found <110>{111}-type 60° dislocations lying on the glide plane which is consistent with our results for h = 10 nm. 60° dislocations have been consistently observed for other fcc metal bilayers such as β-Co/Cu [9], and Pt/Au [15]. Their critical thickness h_c was about 1.46 nm compared to a theoretical prediction of h_c = 1.9 nm. The trend of the dislocation spacing vs. layer thickness in the current study is consistent with their observation. Burgers vectors of the misfit dislocations are also in agreement at least over the thickness range (< 10 nm). Transition of the character of the misfit dislocations with increasing h can be understood by knowing the fact that 90° dislocations accommodate the lattice misfit more efficiently than 60° dislocations. However, the formation mechanisms of the two different types of misfit dislocations with variation of layer thickness will need to be elucidated from the point of nucleation and stability of the dislocations.

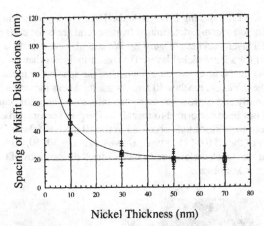

Figure 2. Spacing of misfit dislocations with changing thickness of Ni layer epitaxially deposited on 100 nm Cu layer.

Coherency strains and stresses can be calculated from the measured spacing of misfit dislocations under the suitable assumptions. For the uniformly strained bilayer, biaxial plane-stress may be applied, $\sigma_{11} = \sigma_{22} = \sigma$, $\sigma_{33} = 0$, and $\varepsilon_{11} = \varepsilon_{22} = \varepsilon$ [8, 16, 17]. In the isotropic elasticity, $\sigma = 2\varepsilon\mu(1+\nu)/(1-\nu)$, where μ is the shear modulus and ν is Poisson's ratio. For the bilayer A/B with lattice constants $a_B > a_A$, $|\varepsilon_A| + |\varepsilon_B| + \varepsilon_b = f$, $f = 2(a_B - a_A)/(a_B + a_A)$, $\varepsilon_b = b/\lambda$, total strain $= |\varepsilon_A| + |\varepsilon_B|$, where f is a lattice misfit, b is an edge component on the interface plane of the Burgers vector, and λ is the spacing of the misfit dislocations. With the further assumption that $|\varepsilon_A| = |\varepsilon_B| = \varepsilon$, $\mu_A = \mu_B = \mu$ and $\nu_A = \nu_B = \nu$ for the A and B layers, the strain and stress near the interface are obtained by $\varepsilon = (f - \varepsilon_b)/2$ and $\sigma = 2\varepsilon\mu(1+\nu)/(1-\nu)$. Using average values of Ni and Cu for μ, ν, b, and f ($\mu = 74.7$ GPa, $\nu = 0.30$, $b = 0.2524$ nm, $f = 0.0256$; $a_{Ni} = 0.35238$ nm, $a_{Cu} = 0.36150$ nm, $\mu_{Ni} = 94.7$ GPa, $\mu_{Cu} = 54.6$ GPa, $\nu_{Ni} = 0.276$, $\nu_{Cu} = 0.324$), we can calculate stresses and strains vs. Ni layer thickness (Table 1). In this approximation, the total strain is divided equally between Ni and Cu layers, and therefore is twice the strain magnitude in each layer. At 2 nm thickness, near the critical thickness, the misfit is entirely accommodated by the coherency strains in the two layers and the stress is as high as 3.55 GPa at 2.56 % strain.

Table 1. Coherency stress and total strain calculated from the spacing of misfit dislocations.

Ni thickness (nm)	Total strain	Stress (GPa)
2	0.0256	3.55
10	0.0224	3.11
30	0.0149	2.07
50	0.0127	1.76
70	0.0122	1.69

At 10 nm thickness, the energetically favored formation of misfit dislocations occurs and the coherency stress is reduced to 3.11 GPa at 2.24 % strain. At relatively small thicknesses, 60°

dislocations, which have half of the edge component of 90° dislocations resolved in the interface, form to remove the misfit. As the thickness increases, 90° dislocations, which remove the misfit more efficiently, appear. At this thickness, the misfit dislocation spacing still exceeds the asymptotic value and the dislocations accommodate about half of the misfit strain. The lattice remains strained by 1.22 %, corresponding to a stress of 1.69 GPa, which is higher than the value (~0.7 %) reported by Matthews and Crawford [10]. Their strains were obtained by measuring the moiré fringes caused by misfit averaged over the thickness of the bilayer.

Trilayer

In the trilayer Cu/Ni/Cu, we expect that misfit dislocations form at both interfaces and it is interesting to see how the misfit dislocations are arranged at the interfaces as a result of interaction between them. Figure 3(A) shows a typical example of the misfit dislocations in the trilayer 70 nm Cu/10 nmNi/70 nm Cu. The misfit dislocations form a cross-grid of arrays as for the case of the 10 nm Ni/100 nm Cu bilyer.

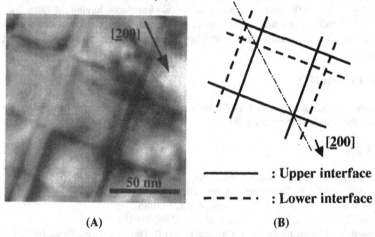

[200]

50 nm

[200]

――――― : Upper interface

－ － － － · : Lower interface

(A)   (B)

Figure 3. Misfit dislocations in the 20 nm Cu/10 nm Ni/70 nm Cu trilayer. The tilting experiment revealed the arrangement of the paired misfit dislocations.

More interestingly, parallel dislocations are often observed in pairs. The paired dislocations are 60° dislocations whose edge components in the interface have opposite signs. In order to identify the configuration of the paired dislocations, a tilting experiment was performed to observe the change of the width of the pairs of dislocations. For the paired dislocation lines crossing in mutually perpendicular directions, the width changes are best observable by tilting along the [200] in the (001) plane. The configuration of the misfit dislocations has been successfully determined. The dislocations designated by the solid lines in Figure 3(B) are located on the upper interfaces, while those designated by the dotted lines are on the lower interface. Evidently, the grid spacing exceeds the layer thickness so that paired dislocations on the top and bottom interfaces assume a near 45° dipole arrangement characteristic of an isolated dipole. Because the dipole-dipole interaction is weak, one expects the dipoles to be oriented nearly randomly between the symmetric configuration in Figure 3 and a symmetric configuration where both top dislocations are at the same side of the bottom dislocations. Here only one case for trilayers was presented, but a study of misfit dislocations in trilayers with different Ni layer thicknesses is planned.

CONCLUSIONS

In the bilayers, the spacing of misfit dislocations depends on the thickness of the Ni layer. The critical thickness is found to be between 2 and 5 nm. The dislocation spacing at the largest thickness is about 20 nm, larger than the expected asymptotic value (~10 nm) corresponding to the lattice misfit. For this case, misfit dislocations accommodated about half of the total misfit. At large Ni overlayer thickness (~30 nm), misfit dislocations are <110>{001} type edge dislocations. At relatively smaller Ni layer thickness (~10 nm), <110>{111} type 60° mixed dislocations are observed lying along the intersection of the {111} slip plane and the {001} interface plane, in both the Ni/Cu bilayer and the Cu/Ni/Cu trilayer films.

In the trilayer, the misfit dislocation arrays at the two interfaces, having the same line directions, are 60° dislocations with opposite edge components. Paired dislocations were formed on the two interface planes and displaced with respect to each other in accordance with expected interaction forces.

ACKNOWLEDGMENTS

The authors would like to thank J. D. Embury and R. M. Dickerson for many valuable discussions in the course of the present study.

REFERENCES

1. F.C. Frank and J.H. van der Merwe, Proc. Roy. Soc. (London), A198, 216 (1949).
2. J.W. Matthews and A.E. Blakeslee, J. Cryst. Growth, 27, 118 (1974).
3. R. People and J.C. Bean, Appl. Phys. Lett., 47, 323 (1985).
4. R. Hull and J.C. Bean, J. Vac. Sci. Technol., A7, 2580 (1989).
5. J. Petruzzello, B.L. Greenberg, D.A. Cammack, and R. Dalby, J. Appl. Phys.,63,2299 (1988).
6. J.S. Ahearn and C. Laird, J. Mater. Sci., 12, 699 (1977).
7. H.-J. Gossman, B.A. Davidson, G.J. Gualtieri, G.P. Schwatz, A.T. Macrander, S.E. Slusky, M.H. Grabow, and W.A. Sunder, J. Appl. Phys., 66, 1687 (1989).
8. J. Narayan, S. Sharan, A.R. Srivasta, and A.S. Nandedkar, B1, 105 (1988).
9. J.W. Matthews, Thin Solid Films, 5, 369 (1970).
10. J.W. Matthews and J.L. Crawford, Thin Solid Films, 5, 187 (1970).
11. K. Shinohara and J.P. Hirth, Thin Solid Films, 16, 345 (1973).
12. J.D. Embury and J.P. Hirth, Acta. Metall. Mater., 42, 2051 (1994).
13. J.W. Matthews, J. Vac. Sci. Tech., 3, 133 (1966).
14. D.M. Mattox, *Handbook of Physical Vapor Deposition (PVD) Processing*, (Noyes Publications, New Jersey, 1998), p. 482.
15. J.W. Matthews and W.A. Jesser, Acta. Metall., 15, 595 (1967).
16. J.P. Hirth and X. Feng, J. Appl. Phys., 67, 3343 (1990).
17. X. Feng and J.P. Hirth, J. Appl. Phys., 72, 1386 (1992).

Mat. Res. Soc. Symp. Proc. Vol. 673 © 2001 Materials Research Society

Structure and mechanical behavior relationship in nano-scaled multilayered materials

A.Sergueeva, N.Mara, A.K.Mukherjee
Department of Chemical Engineering and Materials Science,
University of California, Davis, CA 95616, U.S.A.

ABSTRACT

Multilayered foils with 10%Cu/90%Ni and different bi-layer thickness (100-1000 nm) have been fabricated by electrodeposition. TEM and x-ray diffraction analysis indicate discrete layer formation and a (100) textured structure. The maximum tensile strength (590 MPa) is obtained for foils with the smallest layer thickness. Preliminary results on high temperature deformation show a strong dependence of strength and plasticity on layer thickness.

INTRODUCTION

Multilayered films form a very important group of materials and there is significant interest in studing their fundamental physical properties [1]. More specifically, nanometer-scale polycrystalline multilayered films (with layer thickness less than 100 nm) have been the subject of many recent experimental and theoretical studies [2-5]. These fine-scale composite materials typically exhibit high yield strength, which at room temperature can approach one-half of the theoretical strength. Most attempts to characterize the mechanical behavior of such thin films have been carried out using nanoindentation and scanning force microscopy or their combination and there are no data regarding their mechanical behavior at elevated temperatures. In the present investigation, the microstructure and mechanical properties of polycrystalline Cu-Ni nanolayered composites prepared by electrodeposition were evaluated. Samples were tested in uniaxial tension at room and elevated temperature using a custom-built computer controlled constant strain rate tensile test machine. This allows testing of very small samples with a 1-mm gage length and thicknesses as small as 5 μm, providing new opportunities in mechanical characterization of multilayered materials. The high strength of these new materials is attributed to their layered, nanoscale structure and a variety of related strengthening mechanisms. Variations in composition, stiffness, residual elastic strain, crystal structure and interfacial defects are expected to impede dislocation glide between layers, while intralayer grain boundaries and nanoscale grains restrict dislocation motion within layers. Mechanical testing data suggests that some of these strengthening mechanisms are active in multilayers.

EXPERIMENTAL DETAILS

The 10%Cu-90%Ni multilayers with bi-layer thickness of 100, 300 and 1000 nm have been fabricated by electrodeposition from a single bath. The electrodeposition setup consisted of a rotating disc electrode in 0.5L of Nickel Sulfamate electroplating solution at 50 C. This solution consisted of 76.5 g/L Ni metal in the form of nickel sulfamate, 45 g/L boric acid, 0.15 g/L of Barret SNAP (sulfamate nickel anti-pit), and 5 mM Cu^{2+} added in sulfate form. Multilayer specimens were directly deposited on a rotating cathode of pure titanium polished to a 0.25 micron finish. The anode used was pure nickel. The deposition sequence was as follows: Plate

Cu onto the rotating cathode at a current density of $2 mA/cm^2$, stop rotation for 6 seconds to establish quiescent conditions, then plate Ni at $105 \, mA/cm^2$ under zero rotation. The deposition process was computer controlled using a custom LabView program controlling two independent power supplies and cathode rotation. In this way, 2.5 cm multilayer specimens ranging from 5 to 15 microns thickness were produced. After deposition, tensile specimens were shaped by milling on a CNC hybrid mill or by cutting with a razor blade.

The microstructures of as the processed specimens were investigated by x-ray diffraction method and transmission electron microscopy (TEM). Thin foils for TEM were prepared by microtoming. Differential scanning calorimeter (DSC) measurements were carried out between $25°$ and $550°C$ with a heating rate of $40°C/min$. The tensile tests were performed using a custom-built computer controlled constant strain rate tensile test machine with a displacement resolution of 5 μm and a load resolution of 0.1 N.

RESULTS AND DISCUSSION

The microstructure of layered material with bi-layer thickness of 1000 nm is shown in Fig.1. It is seen that the thickness of the layers is quite uniform but the layers are not straight. One of the reasons of such roughness of layers can be the spherical growth with radial organization and high faceting of fcc Cu [6]. In general, the interfacial roughness should decrease with decreasing layer thickness as it was shown by Moffat [7] for Cu/Co multilayers.

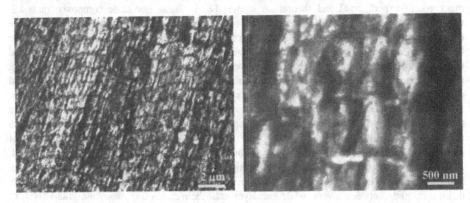

Fig.1. Microstructure of 100 nm Cu/ 900 nm Ni multilayer at different magnifications.

The x-ray diffraction of electrodeposited multilayers exhibits that only the peaks corresponded to pure Cu and Ni present confirming that no intermixing occurs in the as-prepared state (Fig.2). The x-ray diffraction patterns were similar for all investigated multilayers in as-prepared state. Moreover, the results show a predominant (200) peak and a much lower intensity (111) peak. A strong (100) texture associated with Ni dominance is common for multilayers with a large ratio of Ni to Cu thickness [8].

The room temperature tensile testing (Fig.3) revealed a strong dependence of ultimate tensile strength on bi-layer thickness but it showed approximately the same yield stress for multilayers with 300 and 1000 nm wavelength. Decreasing the wavelength to 100 nm results in significant yield stress enhancement. The analysis of data from literature revealed a large scatter in the tensile strength for similarly prepared multilayers. The yield stress of our multilayers with a bi-layer thickness of 100 nm (YS=510 MPa) is not so high as reported by Tench and White [9] (YS=900 MPa) and by Menezes and Anderson [10] (YS=1200 MPa) for the similar Cu/Ni multilayers produced by electrodeposition. However, our data is very close to the data of Embrahimi and Liscamo [8] who reported yield stresses in the range of 376 to 685 MPa for the Ni to Cu ratio of of 9.5. The deposition parameters, which control the evolution of crystallographic texture, the degree of co-deposited Cu in Ni layer, the formation of nodules, etc., seems to be very significant. The comparison of our deposition process with that used by other authors [8-10] shows that the substantially similar parameters were used by Embrahimi and Liscamo [8]. One of the reasons for the high strength reported in [9,10] can be the difference in the amount of Cu during the deposition. Menezes and Anderson [10] showed that increasing co-deposited Cu leads to high strength as a result of increasing coherency of interfaces. Moreover, the interface width and its compositional profile are expected to influence the properties [11].

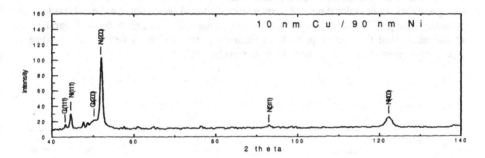

Fig.2. X-ray pattern of electrodeposited 10%Cu-90%Ni multilayer.

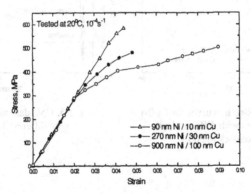

Fig.3. Mechanical properties of multilayers at room temperature.

The most important result of current investigation is concerned with the mechanical behavior of the multilayers at elevated temperature. The stress-strain curves for samples with bi-layer thicknesses of 100 and 1000 nm are shown on Fig.4. DSC measurements (Fig.5) show that there is no significant reaction up to 550°C except the magnetic transformation at about 330°C. According to x-ray diagrams the samples after deformation at 500°C remain multilayered. There are strong peaks corresponded to pure Cu and Ni but with some additional peaks that can be related to the alloy formation because of intermixing. The intensity of such additional peaks increases with decreasing layer thicknesses (Fig.6). It should be noted that the texture of the 10 nm Cu/90 nm Ni sample was changed from (100) to (111) after deformation. There is no data on elevated temperature tensile behavior of Cu/Ni multilayers, so our preliminary results are not enough to make any definitive conclusions, but these first results showed a strong dependence of high temperature deformation on layer thickness. Decreasing bilayer thickness results in high strength, which is a common dependence. At the same time, the sample with smaller layer thickness showed a high ductility. One of the reasons for ductility enhancement can be related to a decrease in the misfit dislocation density with layer thickness. The barrier for slip systems and propagation of dislocations through interfaces can become significantly weaker with increase in the spacing of misfit dislocations. Moreover, the additional elongation can be related to grain boundary sliding within the layers and to interface sliding. However, more detailed investigation of the nature and the role of possible mechanisms is required for a better understanding of macroscopic yield of multilayers with different wavelength.

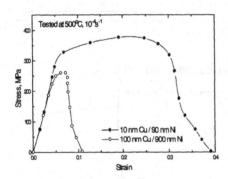

Fig.4. Stress-strain curves for Cu/Ni multilayers with different layer thicknesses at 500°C.

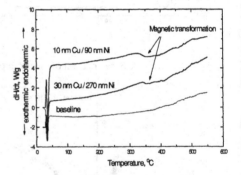

Fig.5. DSC data for Cu/Ni multilayers with different layer thicknesses.

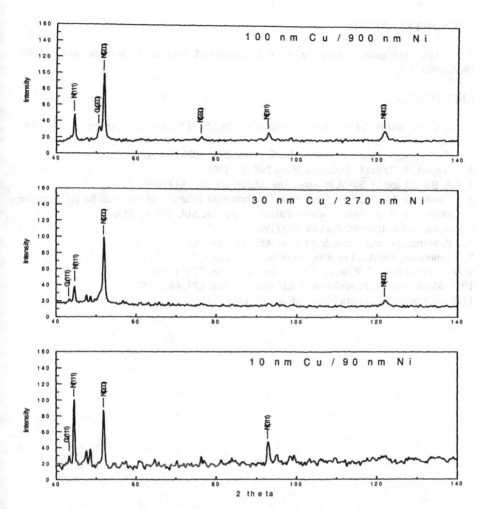

Fig.6. X-ray pattern of multilayers after deformation at elevated temperature.

CONCLUSIONS

10%Cu/90%Ni multilayers with bi-layer wavelength of 100, 300 and 1000 nm have been electodeposited. The room temperature tensile testing (Fig.3) revealed a strong dependence of ultimate strength on bi-layer thickness but for yield stress it showed approximately the same value for multilayers with 300 and 1000 nm wavelength. Decreasing of the wavelength to 100 nm results in significant yield stress enhancement. A strong dependence of high temperature deformation on layer thickness has been demonstrated

ACKNOWLEDGMENTS

This investigation is supported by a grant from U.S. National Science Foundation (NSF-DMR-9903321).

REFERENCES

1. L.L. Chang and B.C. Giessen (eds.), Synthetic Modulated Structures, Acad.Press, New York, 1986.
2. C. Suryananayana and F.H. Froes, Metall. Trans. **23A**, 1071 (1992).
3. S. Veprek, S. Reiprich, Thin Solid Films **268**, 64 (1995).
4. S.A. Barnett and M. Shinn, in Annu. Rev. Mater. Sci. **24**, 481 (1994).
5. J. Grilhe, in Mechanical Properties and deformation behavior of materials having ultra-fine microstructures, (Kluwer Academic Publishers, Boston, MA, 1993) p.255.
6. J. Amblard, Electrochim. Acta **28**, 909 (1983).
7. T.P. Moffat, Mater.Res.Soc.Symp.Proc. **451**, 413 (1997).
8. F. Embrahimi and A.J. Liscamo, Mater. Sci. Eng. **A301**, 23 (2001).
9. D.M. Tench and J.T. White, J. Electrochem. Soc. **138**, 3757 (1991).
10. S. Menezes and D.P. Anderson, J. Electrochem. Soc. **137**, 440 (1990).
11. J.E. Krzanowski, Scripta Metall. Mater. **25**, 1465 (1991).

Mat. Res. Soc. Symp. Proc. Vol. 673 © 2001 Materials Research Society

Dislocations in thin metal films observed with X-ray diffraction

Léon J. Seijbel[1] and Rob Delhez[2]

[1]Netherlands Institute for Metals Research,
Rotterdamseweg 137, 2628 AL Delft, Netherlands
[2]Laboratory for Materials Science, Delft University of Technology,
Rotterdamseweg 137, 2628 AL Delft, Netherlands
email: L.J.Seijbel@tnw.tudelft.nl

ABSTRACT

X-ray diffraction has been used to measure the stress, the crystallite size and the dislocation distribution in thin metal layers. By measuring two orders of a reflection, the contribution of the size distribution to the diffraction line broadening can be eliminated. A model equation is fitted to the strain Fourier coefficients of the diffraction line from which the dislocation arrangement can be obtained. For untextured nickel on steel or on silicon the dislocation densities have been obtained. It is demonstrated that for highly textured layers more information can be obtained than for untextured layers. It was found that a heated molybdenum layer on oxidized silicon showed only inclined screw dislocations.

INTRODUCTION

Thin metallic films are used for many purposes. They almost invariably contain high, internal stresses. These stresses play an important role in the quality of these layers and may even lead to premature failure of the layer. However, the internal stresses may relax by the movement of dislocations.

X-ray diffraction can be used for the determination of dislocations in thin films. Due to the imperfections in the lattice and due to the finite dimensions of the crystallites in the layer, the diffraction lines are broadened. Because dislocations are line shaped crystallographic defects and their motions are related to the crystal structure, the lattice deformation is different for different crystallographic directions. In the XRD-experiment this leads to reflection (HKL) dependent line broadening.

In this paper we present a method to obtain both the dislocation distribution parameters and the average crystallite dimensions. As examples we use three different types of specimen: nickel layers sputtered on a silicon wafer, molybdenum layers sputtered on a silicon wafer, and steel with an electrolytic nickel coating.

THEORY

The volume density E of the elastically stored energy in a crystal is generally written in the form [1]:

$$E = CVGb^2 \rho \ln\left(\frac{R_e}{r_0}\right) \tag{1}$$

where C is a parameter depending on the type of dislocation, V the crystal volume, G the shear modulus, $\mathbf{b}$ the Burgers vector, ρ the dislocation density, r_0 the inner cutoff radius and R_e the outer cut off radius belonging to the dislocation.

When a diffraction line is broadened as a consequence of the presence of dislocations only, then it can be described in terms of R_e and ρ. Denoting A as the Fourier transform of the diffraction profile, it is found that [2]

$$A(L) = A_S(L)A_D(L), \tag{2}$$

with A_S the order independent particle size broadening and A_D describing the order dependent strain (or distortion) broadening. Expressions for these coefficients as function of the correlation distance L are given by [1] and [3]:

$$A_S(L) = \exp\left(\frac{-L}{D_{av}}\right), \tag{3}$$

and

$$\ln(A_D(L)) = -P^g L^2 (Q^g - \ln(L)) \tag{4}$$

with $P^g = \pi/2 \; g^2 b^2 \chi \rho$ and $Q^g = \ln(R_e) + 2\ln(2) - 1/3 - \ln(\mathbf{g} \cdot \mathbf{b} \sin \psi)$, and where $\mathbf{g}$ the diffraction vector, χ the dislocation contrast factor that can be calculated as given by Klimanek and Kuzel [4], ψ the angle between the line vector of the dislocation and the diffraction vector and D_{av} the area weighted average size of the crystallites in the direction of the diffraction vector. The expression for the distortion term Eq.(4) is only valid when the correlation length is smaller then the maximum correlation length L_{max} given by:

$$L_{max} < \frac{1}{2} \frac{R_e}{\mathbf{g} \cdot \mathbf{b} \sin \psi}, \tag{5}$$

which limits the fit range. The dislocation contrast factor χ is a function of the elastic properties of the crystal and is a measure for the microstrain in the crystal in the direction of the diffraction vector so it also depends on the direction of $\mathbf{g}$.

To be able to measure the microstrain in the crystallites, the easiest way is to measure two orders of the same reflection: $\mathbf{g}$ and $n\mathbf{g}$, e.g. {111} and {222} or {200} and {400}. When the ratio of the obtained Fourier coefficients is calculated, the following relation is obtained:

$$\ln\left(\frac{A^{ng}(L)}{A^g(L)}\right) = \ln\left(\frac{A_D^{ng}(L)}{A_D^g(L)}\right) = -P^g L^2 \left((n^2 - 1)(Q^g - \ln L) - n^2 \ln n\right) \tag{6}$$

By fitting P^g and Q^g to the measured Fourier coefficients the dislocation density ρ and the outer cut off radius R_e are obtained. Extra proof for the model is found when the thus obtained Fourier coefficients not only fit the ratio of the two reflections but also fit the individual profiles. This can be checked by calculating the resulting size coefficients of the two measured orders which must be equal for the two measured reflections.

EXPERIMENTAL

Sample preparation

As first example we show measurements performed on 2 μm thick nickel coatings electrodeposited on a steel substrate. These layers have a columnar structure with crystallite sizes of the order of 100 nm. One specimen was measured as it was prepared. Another specimen was annealed for three hours at 525 K. At higher temperatures mixing of substrate and coating starts, which makes the technique presented unusable.

As a second example we show measurements performed on 500 nm thick nickel layers sputtered on an oxidized silicon wafer. TEM images of the layers show that the average crystallite size is of the order of 20 nm. The specimens were annealed for 30 minutes in an Ar/5%H₂ atmosphere at temperatures of 330 K, 350 K, 375 K, 400 K and 450 K respectively and cooled to room temperature fast. One sample was heated at 500 K for two hours after which it was gradually cooled down to room temperature. All diffraction experiments were preformed at room temperature.

The last experiment presented is performed on a 500 nm thick molybdenum layer sputtered on an oxidized silicon wafer. The sample was heated up to 800 K with a heating rate of 2 K/s, held at 800 K for 2 hours after which it was cooled down to room temperature with a cooling rate of 25 K/s. Texture measurements show that most of the crystallites in the layer have a {100} plane parallel to the surface and a <110> direction parallel to tangential direction of the rotating sputter table [5]. Therefore, the number of possible orientations of slip planes in the crystallites is small.

X-ray diffraction

The diffraction lines were measured using two Bruker AXS D5005 X-ray diffractometers. The diffractometer on which the molybdenum specimens were measured, was equipped with a Huber open Eularian cradle, allowing all possible tilt angles Ψ. The XRD-measurements were made using CuKα radiation applying a diffracted beam monochromator. Line broadening and displacement by instrumental errors were corrected using XRD-measurements taken from a nickel powder for the nickel layers and from a ZnO powder for the molybdenum layers. The diffraction lines obtained from these powders show negligible contributions of lattice deformation and finite dimensions of the crystallites to line breadths.

RESULTS AND DISCUSSION

Nickel on steel

In Figure 1 it can be seen that the stress in the nickel coating on steel after annealing 3h at

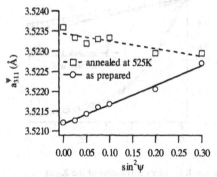
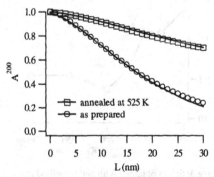

Figure 1: Sin$^2\psi$-plots [5] of the {311} reflection of a nickel coating on steel. After preparing the coating the stress was found to be 210 MPa. After annealing at 525 K for 3 hours the resulting stress becomes -80 MPa.

Figure 2: Measured and fitted Fourier coefficients of the X-ray diffraction profiles of the {200} reflection of a nickel coating on steel as a function of the correlation factor L of the as prepared sample and the sample annealed for 3 hours at 525 K.

525 K becomes negligible. When we want to know what has happened inside the coating we can look at the Fourier coefficients calculated from the diffraction lines of the {200} and the {400} reflection. By fitting Eq. (6) to these results the average crystallite size and the dislocation density can be obtained. Figure 2 shows the fit to the Fourier coefficients of the {200} reflection. To calculate the dislocation density an average contrast factor for screw dislocations was taken. It was found that the dislocation density decreased from 6×10^{14} m^{-2} to 2×10^{14} m^{-2} after annealing. The width of the diffraction lines decreased not only due to a decrease in dislocation density but also due to the growth of the crystallites. In the as prepared sample the average crystallite size was 100 nm. After annealing at 525 K the observed average size becomes 400 nm. The outer cut off radius does not depend on the crystal dimension but rather on the dislocation distribution [6]. This can be concluded from the fact that R_e did not increase, but decreased from 50 nm to 27 nm after the heat treatment. From the values of R_e the maximum correlation length can be obtained for which the model is valid. For the {111} reflection the L_{max} values were found to be 30 and 17 nm for the as prepared sample and the annealed sample respectively. It is clear that in this range the fit is quite good.

Nickel on silicon

The nickel layers on silicon all contain small crystallites. The diffraction line broadening observed consists of size broadening as well as strain broadening as was found after the calculation of the ratio of the Fourier coefficients of the {111} and {222} reflection. Figure 3 shows the stress measured from the shift of the {311} reflection as a function of the tilt angle. The stresses in the layers are relaxed after annealing at a temperature of about 400 K. Due to the difference in thermal expansion of layer and substrate extra stress is induced by annealing. This extra stress becomes relevant when the initial growth stress has been annealed.

The Fourier coefficients of the as prepared layer and the sample annealed at 500 K are shown in Figure 4. From the fitted value of Q^g, the maximum value of the range in which the model is valid could be obtained again. For these two examples L_{max}= 16 and 36 nm for the as prepared and the annealed sample respectively. It is clear that the fit is not valid in that range for

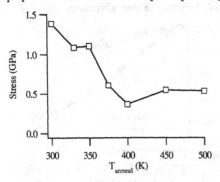

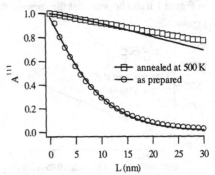

Figure 3: Stress obtained with $\sin^2 \psi$ method in nickel layers on silicon as a function of the anneal temperature. Annealing at various temperatures reduces the internal stress in the layer. Annealing at higher temperatures results in the introduction of thermal stresses.

Figure 4: Fourier coefficients of the X-ray diffraction profiles of the {111} reflection of nickel layers on silicon as a function of the correlation factor L. After annealing the diffraction lines are less broadened as a result of crystallite growth and a reduction in internal strains.

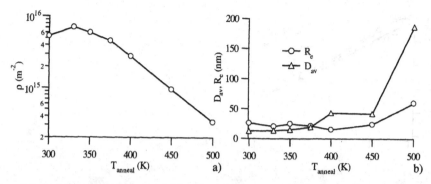

Figure 5: The dislocation density (Figure a) and outer cut off radius (Figure b) of the dislocation distribution in nickel layers on silicon as a function of the anneal temperature. The average crystallite size is also plotted in Figure b.

the annealed sample. This is due to inaccuracies of the measurement.

When the fit is performed for $L \ll L_{max}$, the dislocation density, the crystallite sizes and the outer cut off radii are obtained. Again for the calculation of the dislocation density an average value of the contrast factor for screw dislocations was taken. In Figure 5a the dislocation density as a function of the anneal temperature is given. It is clear that the dislocation density reduces with the internal stress as shown in Figure 3. The increase of the stress due to the difference in thermal expansion does not influence the dislocation density. Figure 5b shows the area weighted average crystallite size D_{av} and the outer cut off radius R_e as a function of the anneal temperature. Whereas the crystallite size clearly increases with increasing temperature, the outer cut off radius does not change significantly.

Molybdenum on silicon

The molybdenum sample was heat treated during 2h at 800 K. When stresses relax by the introduction of dislocations, it is expected that screw dislocations with Burgers vectors inclined to the specimen surface are produced. Since the molybdenum specimen has a strong texture only four orientations of the Burgers vector can be found, two of which are inclined to the specimen surface. When the angle between the diffraction vector and the Burgers vector equals 0° or 90°, the dislocation contrast factor becomes 0. So when the tilt angle $\psi = 0°$, both inclined dislocations can be seen. When $\psi = 60°$ one inclined and one parallel dislocation becomes visible. When there are no dislocations with a parallel Burgers vector the dislocation density measured for the tilted specimen must be half of the dislocation density found without tilting the specimen.

Figure 6 shows the Fourier coefficients of the {110} reflection for zero tilt and for a tilt of 60°. Although the fit is not very good the obtained results are very good. At zero tilt, a dislocation density of 4.60×10^{15} m^{-2} was found, for $\psi = 60°$ a dislocation density of 2.32×10^{15} m^{-2} was found. This is exactly the factor of two that was expected on the basis of the contrast factor and the presence of only dislocations with inclined Burgers vector. The average crystallite size in the direction of the diffraction vector found for these two tilt angles are infinite and 50 nm for respectively $\psi = 0$ and 60°. The value infinite was found since $1/D_{av}$ is fitted. The value of $1/D_{av}$ is too small to measure accurately which means that the columns are of the order of a few hundreds of nanometers. The width of the columns therefore is of the order of 40 nm.

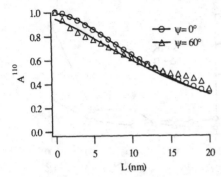

Figure 6: Fourier coefficients of the diffraction profile of the molybdenum specimen seen under two angles. The difference in shape is due to the different dimensions of the crystallites and the difference in dislocation contrast.

CONCLUSIONS

- The Wilkens model to describe dislocations in thin layers observed with X-ray diffraction can be used in various cubic materials with different textures. For specimens with a (sharp) texture, the results are more detailed than for specimens with (almost) random texture.
- The dislocation densities of nickel layers on steel or silicon decrease with the decrease of the growth stress after annealing. The observed dislocations are a consequence of the layer growth. The number of dislocations nucleated upon cooling after annealing the layers is negligible, because the thermal mismatch between Ni and steel is very small.
- In the annealed molybdenum layer the stress is relaxed by the introduction of screw dislocations with Burgers vectors inclined to the plane of the layer. Due to its sharp texture it is possible to distinguish between the possible dislocation types.

ACKNOWLEDGEMENTS

This research was carried out under project number MS97007 in the framework of the Strategic Research program of the Netherlands Institute for Metals Research in the Netherlands (www.nimr.nl). Dr. Emile van der Drift of the Delft Institute of Microelectronics and Submicron Technology DIMES is acknowledged for making available the sputter facility.

REFERENCES

1. M. Wilkens, *Phys. Stat. Sol.* **2**, 359, (1970)
2. B.E. Warren in *X-Ray Diffraction*, (Addison-Wesley, Reading MA, 1969)
3. R. Delhez, Th. H. de Keijser, and E.J. Mittemeijer, *Fresenius Z. Anal. Chem.* **312**, 1, (1982)
4. P. Klimanek and R. Kuzel, *J. Appl. Cryst.* **21**, 59, (1988)
5. I.M. van den Berk, L.J. Seijbel, and R. Delhez, submitted to: *Mechanisms of Surface and Microstructure Evolution in Deposited Films and Film Structures*, (MRS proceedings spring 2001)
6. J.-D. Kamminga and R. Delhez, *Mater. Sci & Engin. A* **309-310**, 55, (2001)

Mat. Res. Soc. Symp. Proc. Vol. 673 © 2001 Materials Research Society

Local Microstructure and Stress in Al(Cu) Thin Film Structures Studied by X-Ray Microdiffraction

B.C. Valek[1], N. Tamura[2], R. Spolenak[3], A.A. MacDowell[2], R.S. Celestre[2], H.A. Padmore[2], J.C. Bravman[1], W.L. Brown[3], B. W. Batterman[2,4] and J. R. Patel[2,4]

[1] *Dept. Materials Science & Engineering, Stanford University, Stanford CA 94305 USA*
[2] *ALS/ LBNL, 1 Cyclotron Road, Berkeley CA 94720 USA*
[3] *Agere Systems, formerly of Bell Laboratories, Lucent Technologies, Murray Hill NJ 07974 USA*
[4] *SSRL/SLAC, Stanford University, P.O.BOX 43459, Stanford CA 94309 USA*

ABSTRACT

The microstructure of materials (grain orientation, grain boundaries, grain size distribution, local strain/stress gradients, defects, ...) is very important in defining the electromigration resistance of interconnect lines in modern integrated circuits. Recently, techniques have been developed for using submicrometer focused white and monochromatic x-ray beams at synchrotrons to obtain local orientation and strain information within individual grains of thin film materials. In this work, we use the x-ray microdiffraction beam line (7.3.3) at the Advanced Light Source to map the orientation and local stress variations in passivated Al(Cu) test structures (width: 0.7, 4.1 μm) as well as in Al(Cu) blanket films. The temperature effects on microstructure and stress were studied in those same structures by *in-situ* orientation and stress mapping during a temperature cycle between 25°C and 345°C. Results show large local variations in the different stress components which significantly depart from their average values obtained by more conventional techniques, yet the average stresses in both cases agree well. Possible reasons for these variations will be discussed.

INTRODUCTION

Extensive study has been conducted on the mechanical properties of thin films and structures. Thermal expansion mismatch between materials and/or transport of material during electromigration lead to large stresses in these structures, often much higher than those sustainable by bulk materials. Conventional techniques such as wafer curvature and x-ray diffraction only provide a macroscopic average of strain/stress or film texture. As integrated circuit device dimensions shrink to submicrometer sizes, one can expect that local microstructural and stress variations do play a prominent role in determining the materials failure modes. X-ray microdiffraction techniques developed at synchrotron sources [1,2,3] have been shown to be a promising new method in the study of mechanical behavior at the micrometer scale. These techniques allow for the determination of the orientation of single grains in a material as well as the complete strain/stress tensors with micrometer scale resolution. X-rays are particularly advantageous because they are penetrating and can probe buried structures. This means that special sample preparation is unnecessary, allowing the investigation of passivated interconnect structures without altering the stress state.

EXPERIMENTAL

This study was conducted at the x-ray microdiffraction beam line (7.3.3) at the Advanced Light Source. A more detailed description of the beam line is given elsewhere [4]. White x-rays (6 keV – 14 keV) are focused down to a 0.8 x 0.8 micrometer spot using a specially configured Kirkpatrick-Baez mirror pair. Because the beam spot is so small, rotation of the sample is not allowed, as the beam would not stay at the same position on the sample. With the sample fixed we use a white x-ray beam to obtain Laue patterns from individual crystallites. A piezoelectric stage allows for precision positioning of the sample in the focal plane of the x-rays. A 4K x 4K SMART 6000 Bruker CCD collects the Laue pattern at each position on the sample as it is translated. Each CCD frame may contain multiple Laue patterns from different metal grains and the silicon substrate. By translating the sample in this manner, we obtain a series of Laue patterns, which enables us to create maps in stress, strain and orientation.

Automated software allows for rapid processing of the CCD frames. The Si background pattern is digitally removed and the remaining Laue spots are indexed. The software can index overlapping Laue patterns in the same frame and recognize Laue patterns which appear in more than one frame as belonging to the same grain. Using the intensities of the Laue patterns in each frame, a map of the grain structure can be produced. The indexing provides the orientation matrix of each individual grain under the submicrometer illuminated area. The misorientation between any two grains as well as orientation variations within single grains can easily be determined with a precision of $0.01°$. The deviations of the Laue spots position from an unstrained crystal are used to calculate the distortion of the crystal unit cell, giving the deviatoric strain tensor with an accuracy of about 2×10^{-4}. Knowledge of the unstrained lattice parameter is unnecessary. The deviatoric stress tensor is simply found by using the anisotropic stiffness coefficients for the material. In this manner, we know the complete orientation and deviatoric stress/strain tensor for each position on the sample.

The samples investigated here are sputtered Al(0.5 wt.% Cu) thin film structures. A 100x100 μm bond pad on the chip (with a thin Ti underlayer) is used to simulate a blanket film (unpassivated, except for the native oxide). The patterned lines have dimensions 0.7 or 4.1 μm in width, 30 μm in length and 0.75 μm in thickness. There are two shunt layers of Ti at the bottom and the top of the lines (thicknesses are 450 Å and 100 Å respectively). The lines are passivated with 0.7 μm of SiO_2 (PETEOS).

The bond pad (blanket film) was thermally cycled between 25°C and 345°C in 40° steps. At each temperature increment, a 15x15 μm area of the film was scanned with the focused white x-ray beam in 1 μm steps. Each scan took approximately 1.5 hours. A 0.7 μm wide line and 4.1 μm wide line were scanned in 0.5 μm intervals at room temperature. In addition, a 0.7 μm line was mapped in 0.5 μm steps across the line and 1 μm steps along the line at several temperatures during a cycle between 25°C and 305°C. The x axis is considered to be along the line, the y across, and the z is out of plane.

RESULTS AND DISCUSSION
Blanket Film

Orientation mapping of the 15x15 μm section of the Al(Cu) bond pad reveals a strong (111) out of plane texture and a random in-plane texture. The out of plane texture ranges from 0 – 3.5° from the sample normal. Grain size ranges from 2.5 µm in diameter to less than 0.5 µm, as revealed by a fine 0.5 µm step scan on a 5x5 µm area of the pad. Due to the limited number of grains illuminated by such small scans, we cannot obtain any accurate statistics on the grain size distribution.

X-ray stress measurement methods which average over a large area assume that the surface of the film is unconstrained and can support no stress ($\sigma_z = 0$)[5]. In order to compare our experimental data, we calculate the average biaxial stress in the film by using the equation:

$$\sigma_{biaxial} = \frac{\left(\sigma'_x + \sigma'_y\right)}{2} - \sigma'_z$$

where σ'_x, σ'_y, and σ'_z are the average deviatoric stress components in the film. Note that σ'_z is equal in that case to the negative of the hydrostatic stress. A plot of the average biaxial stress in the film obtained by x-ray microdiffraction during a thermal cycle between 25°C and 345°C is shown in Figure 1. This behavior is very similar to that seen by Venkatraman *et al.* [6] who performed wafer curvature measurements on Al(Cu) films, including the stress drop after initial plastic deformation during heating and the increase in slope under 200°C upon cooling.

At the local level σ_z generally cannot considered to be equal to zero, especially near grain boundaries. Our experimental data shows that σ_x' is in general not equal to σ_y', indicating that the stress is only biaxial on average. The complex interaction between neighboring grains induces local triaxial stresses. Because the trace of the deviatoric stress tensor must be zero, the z component is an indication of the variation in the x and y components (in-plane stress). Figure 2 shows a map of σ_z' on the bond pad after completion of the thermal cycle. There are clearly large variations in the local deviatoric stresses.

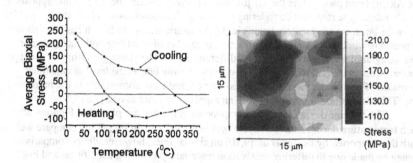

Figure 1: Average biaxial stress vs. temperature for a 15 µm square area of the Al(Cu) film during a thermal cycle.

Figure 2: Map of Z component of deviatoric stress for a 15 µm square area of the Al(cu) film at 25 °C.

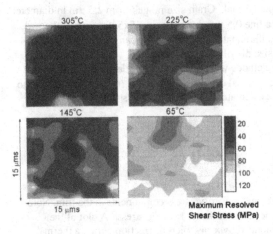

The combination of the deviatoric stress tensor and the complete orientation matrix for a given grain allows us to extract the maximum resolved shear stress (MRSS) for a given slip system. The shear stress is of particular interest when dislocation glide is the dominant deformation mechanism. A map of the resolved shear stress therefore immediately displays where the material is likely to yield first. Figure 3 presents maps of the maximum resolved shear stress on the {111} planes in the <110> type directions (12 independent slip systems) while cooling the film from 305°C to 65°C. The average MRSS at 305°C is 35.5 MPa and increases to 87.5 MPa at 65°C. The average MRSS at 225°C and 145°C are 57.2 and 59.8 MPa, respectively, which means there is very little hardening of the film as it cools over this temperature range. This is clearly visible in Fig. 1. There is a wide range in the MRSS locally at each temperature. Note that these maps cannot be compared with each other directly because each is from a slightly different area on the bond pad, due to sample drift after each temperature change.

Figure 3: Maximum resolved shear stress for a 15 μm square area of an Al(Cu) film at four temperatures during cooling.

4.1 and 0.7 μm Wide Passivated Lines

Grain orientation and stress maps were taken at room temperature for a 4.1 μm and a 0.7 μm wide passivated Al(Cu) line. The grain orientations are similar to those in the Al(Cu) bond pad. While the 4.1 μm line is polycrystalline, the 0.7 μm line appears to be a bamboo-type structure considering the size of the microbeam. In both the 4.1 and 0.7 μm wide lines, we see local variations in the deviatoric stress. In fact, it has been shown in previous work that x-ray microdiffraction is capable of resolving both intergranular and intragranular orientation and stress differences [3]. The stress goes from an average biaxial stress in the pad to a triaxial stress in the narrow line as the level of constraint increases. The average value of the stress along y and z are comparable in the narrowest line. This observation is consistent with an aspect ratio close to one [7,8].

The average deviatoric stress components in the line are plotted in Figure 4 at each temperature during a thermal cycle between 25°C and 305°C. They compare well with those reported by Besser *et al.* [9,10] on similar lines, however, direct comparison cannot be made due to differing passivation materials, processing conditions, and line aspect ratios. The hysteresis in the x and z components indicates some plastic deformation occuring in the line during the thermal cycle. The plastic deformation is greatly reduced compared with the plastic deformation that occurs in the blanket film.

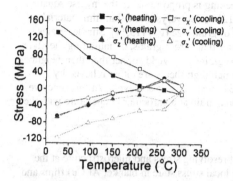

Figure 4: Deviatoric stress components in an Al(Cu) line during a thermal cycle.

This can be attributed to the passivation, which constrains the line on the surface and sidewalls, limiting dislocation motion. The fact that the line is narrow also reduces the average grain size. The effect of passivation constraint on dislocation motion in lines has been described by others [6,11]. Again, the maximum shear stress resolved for the {111}<110> slip system is of particular interest. Figure 5 is a plot of the maximum RSS along the line at three different temperatures when cooling from 150°C to 25°C, with the in-plane and out of plane orientations of the grains along the length of the line. Notice that there are large changes in the shear stress along the line during cooling. The average MRSS increases from 71.2 MPa at 150°C to 112.2 MPa at 25°C as the sample is cooled. By comparison, the average MRSS in the pad at 25°C after the thermal cycle is 95.5 MPa.

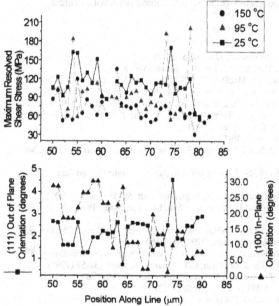

The local variations in stress seen in the blanket film and line at a given temperature can be explained by microstructural mechanisms. First, the combination of misorientation between grains and the elastic anisotropy of the crystal lattice will lead to stress variations even if the strain is constant across the grains. This effect is expected to be minimimal due to the (111) texture of the Al(Cu) grains and the small elastic anisotropy of aluminum. The second mechanism is probably due to the grain size distribution.

Figure 5: Maximum resolved shear stress and grain orientation along the length of an Al(Cu) passivated line. The shear stress is plotted for three temperatures while cooling the line from 305 °C.

It is well know that grain boundary strengthening is proportional to the inverse square root of grain size ($g^{-1/2}$) for bulk materials [12]. In polycrystalline thin film materials, the yield stress is proportional to the reciprocal of grain size (g^{-1}) [13,14]. Smaller grains should be able to support higher shear stresses than larger grains, by inhibiting the glide motion of dislocations. Consequently, some grains may yield sooner than their neighbors and lead to a very complex stress state in which high shears are either relieved by yielding or supported until a critical resolved shear stress is achieved. At this time, we have not correlated the shear stress values in a grain at a particular temperature with the grain size.

CONCLUSIONS

X-ray microdiffraction is capable of resolving stress and microstructure at the micrometer scale. We have shown that the local stress state in blanket Al(Cu) films and in passivated Al(Cu) lines is extremely complex. There are large variations in stress between grains, as well as variations within single grains. The stress state of a blanket film can be on average biaxial, but locally the stress is triaxial. Trends in the average data obtained with x-ray microdiffraction agree well with previous studies of stress in thin film structures. Variations in the local stress (on a micrometer scale) can be attributed to a combination of elastic anisotropy, grain misorientations, and grain size effects. Future work will concentrate on following the variation in stress in single grains as a function of misorientation and grain size, as well as interactions with nearest neighbors.

REFERENCES

[1] J.-S. Chung, and G. E. Ice, J. Appl. Physics, **86**, 5249-5255 (1999)
[2] K.J. Hwang, G.S. Cargill III, and T. Marieb, Mat. Res. Soc. Symp. Proc. **612** (2000)
[3] N. Tamura, B.C. Valek, R. Spolenak, A.A. MacDowell, R.S. Celestre, H.A. Padmore, W. L. Brown, T. Marieb, J.C. Bravman, B.W. Batterman, and J.R. Patel, Mat. Res. Soc. Symp. Proc. **612** (2000)
[4] A.A. MacDowell, R.S. Celestre, N. Tamura, R. Spolenak, B.C. Valek, H.A. Padmore, W. L. Brown, T. Marieb, J.C. Bravman, B.W. Batterman, and J.R. Patel, Proceeding of the 7th International Conference on Synchrotron Radiation Instrumentation, in press (2000)
[5] J.A. Bain, B.M. Clemens, MRS Bulletin 17, 46-51 (1992)
[6] R. Venkatraman, J.C. Bravman, W.D. Nix, P.W. Davies, P.A. Flinn, D.B. Fraser, Jour. of Elect. Mat. **19**, 1231-1237 (1990)
[7] T. Hosoda, H. Yagi, and H. Tsuchikawa, 1989 International Reliability Physics Symposium Proceedings, IEEE, 202-206 (1989)
[8] B. Greenbaum, A.I. Sauter, P.A. Flinn, and W.D. Nix, Appl. Phys. Lett. **58**, 1845-1847 (1991)
[9] P.R. Besser, A. Sauter Mack, D. Fraser and J.C. Bravman, Mat. Res. Soc. Symp. Proc. **309**, 287-292 (1993)
[10] P.R. Besser, *Stress Induced Phenomena in Metallization,* 5th International Workshop, Editors O. Kraft, E. Arzt, C.A. Volkert, P.S. Ho, and H. Okabayashi, 229-239 (1999)
[11] D. Jawarani, M. Fernandes, H. Kawasaki, P.S. Ho, *Stress Induced Phenomena in Metallization,* 3rd International Workshop, Editors P.S. Ho, J. Bravman, C.Y. Li, and J. Sanchez, 32-57 (1995)
[12] R.W. Hertzberg, *Deformation and Fracture Mechanics of Materials,* 117-118 (1989)
[13] R. Venkatraman and J.C. Bravman, J. Mater. Res. **7**, 2040 (1992)
[14] C. V. Thompson, J. Mater. Res. **8** (2), 237-238 (1993)

The Advanced Light Source is supported by the Director, Office of Science, Office of Basic Energy Sciences, Materials Sciences Division, of the U.S. Department of Energy under Contract No. DE-AC03-76SF00098 at Lawrence Berkeley National Laboratory.

Mat. Res. Soc. Symp. Proc. Vol. 673 © 2001 Materials Research Society

Deformation Microstructure of Cold Gas Sprayed Coatings

C.Borchers, T.Stoltenhoff, F.Gärtner, H.Kreye, H.Assadi[1]
University of the Federal Armed Forces, Dept. Maschinenbau, Hamburg, Germany
[1] Dept. of Materials Science Engineering, Tarbiat Modarres University, Tehran, Iran

ABSTRACT

Cold Gas Spraying is a new coating technique, in which the formation of dense, tightly bonded coatings occurs only due to the kinetic energy of high velocity particles of the spray powder. These particles are still in the solid state as they impinge on the substrate. Adiabatic heating after impingement can cause local shear instabilities and jet formation. The local microstructure is strongly dependant on local stress state and temperature rise. A variety of different microstructures is observed by TEM. The results are compared with modelling of the spray process.

INTRODUCTION

In conventional thermal spray techniques like Flame Spraying, High Velocity Oxy-Fuel Spraying or Plasma Spraying, the spray process is accompanied by a partial melting of the powder feedstock material, which can result in high residual stresses of the coatings due to the shrinkage during rapid solidification on the substrate, and oxidization of the material where the process takes place in ambient atmosphere. The former can detract the adhesion and contiguity of the coating, whereas the latter has direct influence on phase formation because of oxides at particle-particle interfaces and therefore determines the physical properties of the coating, e.g. its resistivity.

In the new coating technique of cold gas spraying (CGS), bonding of particles is a result of extensive plastic deformation and related phenomena at the interface [1,2,3]. In the process, solid copper particles in size range between 5 and 25 μm are introduced into the high pressure chamber of a converging-diverging Laval type nozzle and are accelerated in a supersonic stream by a propellant inert gas, which has been preheated to temperatures below 600°C. The particles reach velocities of 500 to 1000 m/s. Because of the low thermal influence, the method can open up new applications for metastable, nanocrystalline or oxidation sensitive materials.

On one hand emphasis of this work is given to the modelling of particle impact where special attention is given to the criteria for bonding and here mainly to a critical particle velocity that the particles must not fall below. In addition the influence of impact angles, and the possible formation of regions where viscous flow of the material occurs are investigated. The modelling presented here is done for the example of copper. On the other hand, the deformation microstructure and binding mechanism of copper coatings are examined. The investigations presented in this work are performed by transmission electron microscopy (TEM) as well as resistivity measurements.

EXPERIMENTAL

Cold sprayed copper coatings are prepared by using nitrogen as process and carrier gas. To obtain a maximum deposition efficiency and dense coatings, preheating of the process gas to 653 K and a gas inlet pressure of 2.5×10^6 Pa were required. As feedstock a gas atomised, 99.8 % pure Cu-powder with an oxygen content of less than 0.1 wt.-% in the size range between 5 and 22 μm was used. For copper coatings, a minimum particle velocity of 570 m/s is necessary to obtain dense, tightly bonded coatings. More general aspects of the coating technique and the determination of a critical velocity are reported in [1,2,3]

Dynamic deformation of the particles upon impact was modelled using ABAQUS, a generalized finite-element software package. The analysis accounts for strain hardening, strain rate hardening, heat dissipation due to plastic deformation, and thermal softening. The heating is assumed to be adiabatic. Most of the impact energy will result in temperature rise, but in order to leave a margin for heat conduction and stored energy, only 90 % of the plastic dissipation is assumed to lead to a temperature rise in the modelling.

To investigate the sprayed coatings by transmission electron microscopy (TEM), cross sections of layers on their respective substrates were sectioned, then discs of diameter 3 mm punched. These were polished, dimpled, and ion milled. Transmission electron microscopy was performed with a Philips EM 420ST.

For resistivity measurements, plates with thickness 150 to 450 μm were cut parallel to the substrate. Resistivity measurements were performed with a four-point device.

RESULTS

The main results of the modelling are summarized Fig. 1. At comparatively low particle impact velocities v, modelled for $v = 300$ m/s, both strain and flow stress show high values at the particle surface. For comparatively high velocities of $v = 900$ m/s, the modelling reveals high plastic strain but flow stresses falling to zero at the surface, indicating an adiabatic shear instability. This shear instability can be assumed to occur when the local temperature reaches the vicinity of the melting point. By calculating the local temperature at the interface for various impact velocities, it has been found that the melting temperature of copper corresponds to an impact velocity $v = 545$ m/s. The modelling also shows that an adiabatic shear instability can be accompanied by a very rapid flow of material at the interface and result in a fast travelling jet.

In addition, the modelling could show that the deviation of impact angle from the perpendicular direction causes higher shear forces due to friction, leading to a higher temperature rise upon impingement compared to perpendicular impact. The highest temperature rise was found for an impact angle of about 45°.

The feedstock powder was examined by TEM (not shown) and revealed quite high dislocation densities of about 10^{12} /cm^2. In addition, TEM could show that at least smaller particles are single crystals, so the average grain size is well above 5 μm. No oxide shell could be detected, and the shapes of the particles are nearly spherical.

Figure 2(a) illustrates the main features of a cold sprayed copper film. The micrograph shows a triple point of three particles. The particle – particle interfaces are indicated by arrows. The hole in the middle of the micrograph is a result of ion milling. The interfaces

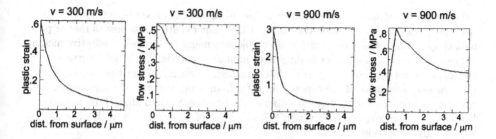

Figure 1: Calculated variation of strain and flow stress with distance from surface of a copper particle with velocities of 300 m/s (left) and 900 m/s (right), 5 ns after the impact. For $v = 300$ m/s, both strain and flow stress are maximum at the surface. For $v = 900$ m/s, the flow stress falls to zero at the surface, indicating a shear instability

themselves are dense. A survey of about 100 TEM-micrographs reveals a porosity of the interfaces of less than 1 %, and a decoration with Cu_2O particles of the same order. In the region of the micrograph marked A, there are equiaxed grains about 100 nm in diameter. In the region marked B, there are aligned elongated grains sized about (50×250) nm^2 in projection. The region marked C seems to be recrystallized with a grain size above 1 μm. In this region, no dislocations can be seen, but there are twins. The region marked D shows a high dislocation density, and again a grain size above 1 μm. In that area the dislocations are piled up in walls. The nanograins in regions A and B also exhibit a distinct interior contrast due to either a high dislocation density or high interior strains.

Figure 2(b) shows a close view of a particle – particle interface and indicates a layer with a thickness of 5 – 15 nm between the particles that appears not to correspond to the microstructures of surrounding particles. The fact that the contrast of this material is identical to the contrast of the particles suggests that this material is as well copper and no oxide.

It should be stressed that although the features shown in fig. 2 are typical for the copper coatings, it is exceptional that they appear in the same micrograph.

The room temperature resistivity $\rho(300K)$ of the cold sprayed copper coating was determined to be 1.7 ± 0.3 $\mu\Omega$cm. The margin of error mainly stems from the uncertainty in defined geometries.

DISCUSSION

The predictions of the modelling were tested by experimental investigations. For the fabrication of dense copper coatings impact velocities above 570 m/s are required [3]. This agrees well with the critical impact velocity of 545 m/s found in the computer experiments on mechanical impact. This seems to confirm that adiabatic strain instability is a prerequisite for successfull bonding, as is the case for explosive welding, where a fast travelling jet of material is observed [4,5]. It is not so easy to certify jet formation in CGS, but the narrow band of metallic material between two particles, see Fig. 2(b), is a strong indicator for jet occurence. The metallic nature of these bands is confirmed by the fact that the room temperature resistivity of the CGS copper coating is as low as that of pure bulk copper, which has a value of 1.7 $\mu\Omega$cm [6]. Jets might be formed by selective melting of near surface regions of either impinging or impinged material. It should be stressed, though, that not all particle – particle interfaces exhibit such jet formation. Apart from this, the microstructure of the copper film is far from being uniform. On the one hand, there is the formation of nanograins either equiaxed or oblong and aligned, on the other hand there are micrograins either with twins and free of dislocations or without twins but with a high dislocation density.

In regions A and B of Fig. 2, nanograins have been formed, but obviously by different mechanisms. In explosive welding, equiaxed nanograins are formed in a binding zone that has a thickness of one micron [5]. Here, the high shear stress leads to a local temperature rise above the melting temperature with subsequent rapid solidification of the liquified material. The resulting microstructure consists of nanograins with a size of about 100 nm, similar to those in region A in Fig. 2. The interface where this is observed is indeed inclined about 45° to the flight direction, the angle for which the modelling predicted a maximum temperature rise. Aligned grains are typical for very high shear stresses but temperatures somewhat beneath the melting point. Elongated grains as in region B of Fig. 2 can be observed in the high deformation region next to the bonding zone in explosive welding [5]. Recrystallization is observed in region C in Fig. 2, characterized by twins and the absence of dislocations. Grains developed by secondary recrystallization in cold sprayed copper coatings are typically on the order of several μm in size. The local temperature can be estimated to be above 600°C [7,8]. In region D in Fig. 2, the microstructure resembles that of the feedstock powder. The observation demonstrates mainly the migration and possible formation of dislocations. The local temperature rise can be expected to be least in that region.

The observed local microstructures seem to be governed by the local stress and temperature rise. Where the temperature rise is highest, local melting and subsequent rapid solidification will occur, leading to equiaxed nanograins. Where the temperature remains below the melting point, but shear stress is very high, oblong aligned nanograins are observed. At temperatures above $\sim$ 600°C, secondary recrystallization with conspicuous twinning can occur. Finally, where the local temperature stays moderate and the stresses are significantly lower than in localized surface regions, only minor changes in microstructure compared to the feedstock powder takes place. In agreement to the computer experiments the local microstructure appears to depend on the angle of particle flight directions to the substrate. In areas of low deformation rates the initial powder microstructure is retained.

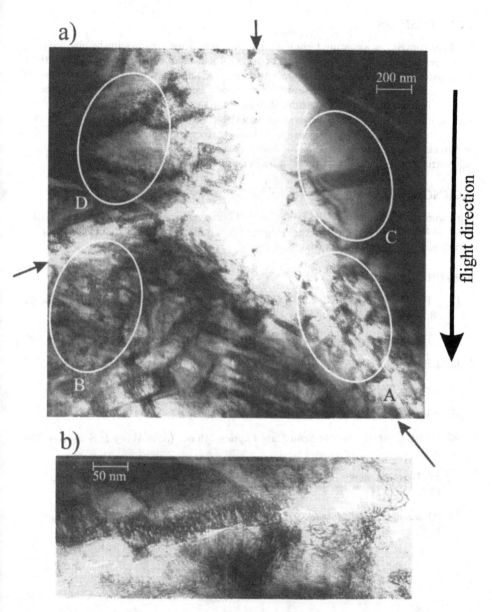

Figure 2: TEM micrograph of a cold sprayed copper coating. (a): Particle – particle interfaces are marked with arrows. The main features are: A: equiaxed grains about 100 nm in diameter. B: aligned elongated grains sized about 50×250 nm in projection. C: recrystallized region, grain size > 1 μm, with recrystallization twins. D: region with high dislocation density, grain size > 1 μm. The dislocations are piled up in walls. (b): Particle – particle interface of a copper film showing a layer with a thickness of $10 - 15$ nm that "belongs" to neither of the particles.

CONCLUSIONS

Cold gas spraying is a coating technique that produces dense coatings in the absence of pores and oxides, which in the case of copper results in room temperature conductivities of bulk material. The observed local microstructures are influenced by local high stresses and fast temperatures rises. In agreement with the modelling of the spray process, local melting can be observed in form of fine solidified equiaxed grains, and interface insertions presumably due to a fast travelling jet of material. The microstructure found in a given region depends on particle size, the angle of impingement, and on the microstructure of the feedstock powder. Therefore on close view to particle – particle interfaces the microstructure of the coatings reveal a quite non-uniform appearance.

ACKNOWLEDGEMENTS

This work was supported by the Deutsche Forschungsgemeinschaft by the grant no. KR 1109 /3-1, which is gratefully acknowledged. HA is grateful to GKSS Research Center, Geesthacht, Germany for a travel grant.

REFERENCES

[1] A.P.Alkhminov, S.V.Klinkov, V.F.Kosarev and A.N.Papyrin, *J.appl.mech.tech.phys.* **38**, 324 (1997)

[2] A.P.Alkhminov, V.F.Kosarev and A.N.Papyrin, *J.appl.mech.tech.phys.* **39**, 318 (1998)

[3] T.Stoltenhoff, H.Richter, H.Kreye, submitted to *J.Thermal Spray Techn.*

[4] H.Kreye, *Wdg.J.Res.Suppl.* **56**, 154 (1977)

[5] H.Kreye, I.Wittkamp and U.Richter *Z.Metallkunde* **67**, 141 (1976)

[6] C.Kittel, *Introduction to Solid State Physics*, 5th ed. (John Wiley & Sons, New York, 1976)

[7] J.G.Byrne *Recovery, Recrystallization and Grain Growth* (MacMillan, New York 1965), p.66

[8] P.Haasen *Physical Metallurgy*, 3rd ed. (Cambridge University Press, Cambridge 1996), p.392

Mat. Res. Soc. Symp. Proc. Vol. 673 © 2001 Materials Research Society

Plastic Deformation of Thin Metal Foils without Dislocations and Formation of Point Defects and Point Defect Clusters

Michio Kiritani, Kazufumi Yasunaga, Yoshitaka Matsukawa and Masao Komatsu
Academic Frontier Research Center for Ultra-high Speed Plastic Deformation,
Hiroshima Institute of Technology, Miyake 2-1-1, Saeki-ku, Hiroshima 731-5391, Japan

ABSTRACT

Evidence for plastic deformation of crystalline metal thin foils without dislocations is presented. Direct observation during deformation under an electron microscope confirmed the absence of the operation of dislocations even for heavy deformation. In fcc metals including aluminum, deformation leads to the formation of an anomalously high density of vacancy clusters, in the form of stacking fault tetrahedra. The dependency of vacancy cluster formation on temperature and deformation speed indicates that the clusters are formed by the aggregation of deformation-induced vacancies. Conditions required for the absence of the dislocation mechanism are explained, and a new atomistic model for plastic deformation of crystalline metals is proposed.

INTRODUCTION

When thin films were pulled out from bulk material by plastic deformation until fracture, the production of an anomalously high density of vacancy clusters was discovered [1]. Other than these point defect clusters, no dislocations were observed in the vicinity. The authors have proposed that the plastic deformation of crystalline metals proceeds by a mechanism that is entirely different from the widely accepted dislocation mechanism. The proposal has not been readily accepted by the research community that has been deeply involved in studying the dislocation mechanism [2]. In the present paper, the authors present evidences for deformation without involving dislocations, and report a variety of experimental observations of the production and annihilation of point defect clusters, providing useful information for establishing a new model of plastic deformation.

THINNING PROCESS OF DUCTILE METAL THIN FILMS BY ELONGATION

Annealed metal films (Al, Au, Cu and Ni) of ribbon shape (20 μm thick, 3 mm wide, and 10 mm long) were cut halfway in the lengthwise central region in order to initiate deformation at that position. The elongation speed of the films varied over a wide range; $10^{-9} - 1$ m/s.

Morphology of the Thinning and Fracture: Localized deformation initiates at the notch-cut, and that position of the film becomes progressively thinner and finally the two sides separate, leaving very thin areas at their tips. These areas are sufficiently thin for observation by transmission electron microscopy [3]. Morphology of the fracture was investigated in detail by Wilsdorf [4], and two typical cases are shown below as the required introduction of this paper.

The fractured tips often exhibit sawtooth shapes as shown in Fig. 1 (a). The fracture often occurs along the trace of the (100) plane as illustrated by AA' in the figure, and it cannot propagate through the long channel of the thinned part when its direction is not parallel to the thinned channel. Consequently, other cracks appear at nearby positions. In order to accommodate the strain between adjacent parts, twinning deformation occurs along the trace of the (111) plane as denoted by B. When this twinning starts, no additional plastic deformation is in progress on either side of the twin plane. The important heavy deformation that is of interest in this paper occurs in the stage before

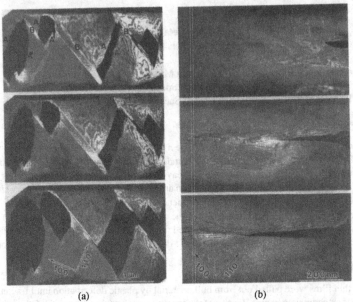

(a)   (b)

Fig. 1. Processes to develop sawtooth-shaped fractured tip (a), and fracture parallel
to a narrow, thinned channel (b).   The case of gold at room temperature.

initiation of cracks. When the direction of a crack, which starts by extreme thinning, is parallel to
the narrow thinned channel, the crack propagates a long distance, as shown in Fig. 1 (b). Typical
elongation direction in this case is along the [111] direction.

Deformation Speed of a Very Local Part: High-speed deformation was performed by simply
pulling apart the ribbon at a speed of about 1 m/s. Slow deformation was performed by use of a
deformation rig whose cross-head speed could be set as low as 1 mm/week. The deformation
presently of interest proceeds on a very narrow band about 1 μm wide, and the deformation speed in
units of strain rate in the two extreme cases mentioned above are 10^6/s and 10^{-4}/s, respectively.   All
the strain rates mentioned in this paper are for this very local part.

DEFECT STRUCTURES INTRODUCED IN ELONGATED THIN FILMS

All deformations in this section were at room temperature, and deformation speed was in the
higher speed range of $10^5 - 10^6$/s in local strain rate. All observations were made with electrons of
sub-threshold energy (120 keV for Al and 200 keV for Au and Cu), avoiding structural change by
radiation damage, and weak beam dark-field imaging condition was adopted with the diffraction
vector of [200], though it is not indicated in each figure.

Gold: As shown in Fig. 2, a very high density of small defects was observed in an area of the tip of
saw-teeth. Dark field weak beam observation along the [110] direction yields clear triangular
images of small defects with their sharp corners pointing in opposite directions as shown in enlarged
micrograph of Fig. 3 (a), indicating that the small defects have the structure of stacking fault
tetrahedra (SFTs) of vacancy type [5]. Average size is about 2.5 nm, and the number density in this

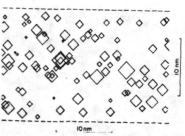

Fig. 2. High-density vacancy clusters formed in an elongated thin part of gold.

Fig. 4. Depth distribution of stacking fault tetrahedra in gold foil, as determined from stereomicroscopy.

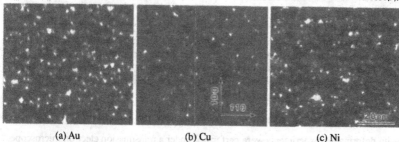

(a) Au   (b) Cu   (c) Ni

Fig. 3. Stacking fault tetrahedra formed in fcc metals by the deformation of thin films to fracture Deformation was at room temperature. Ni was annealed at 573 K after deformation.

case is $1 \times 10^{24}/m^3$. The concentration of vacancies contained in SFTs is estimated to be about 1×10^{-3}. The depth position of each SFT can be determined from the parallax between pairs of stereo-micrographs, and Fig. 4 shows the distribution in the central area of Fig. 2. SFTs are distributed all over the specimen foil, not being strongly influenced by the presence of specimen surfaces.

Copper: The formation of SFTs in copper is the same as in gold as shown in Fig. 3 (b). Only difference is in their smaller size.

Nickel: The point defect clusters formed in nickel at room temperature are very small. After post-deformation annealing at 573 K and above, they grow as shown in Fig. 3 (c). A noteworthy phenomenon is that vacancies in nickel are not mobile at room temperature [6], whereas vacancies in gold and copper do have some mobility at room temperature. The growth of vacancy clusters by annealing in nickel shows that vacancy clusters are not formed directly by deformations but are formed by the clustering of dispersed vacancies introduced by deformation.

Aluminum: Aluminum is listed last because its result was astonishing. Point defect clusters in aluminum are also SFT of vacancy type, as shown in Fig. 5. In aluminum, the formation of SFTs has never been induced by any conceivable experimental treatment; such as quenching from high temperature, irradiation with electrons or heavy energetic particles. Here, the present result shows the inadequacy of the energy consideration of the final nature of clusters.

Fig. 5. Stacking fault tetrahedra formed in aluminum at room temperature.

Fig. 6. Dislocations are observed in the thicker part, but not in the thinner part.

Dislocation Structures: All observations cited in this section were performed under electron diffraction conditions that would have revealed dislocations if any had existed, except for the case of the invisible condition gb=0. However, no dislocations were observed in areas with a high density of vacancy clusters in any of the metals examined. In the border region between the area that is too thick for observation and the thin area with high density of vacancy clusters, complex dislocation structures are generally observed as shown in Fig. 6.

DIRECT OBSERVATION OF DEFORMATION OF THIN FILMS TO FRACTURE

In-situ deformation experiments were performed under a transmission electron microscope. Elongation speed was typically 0.1 μm/s, with the strain rate of the local part being 0.1/s. By the time a sample became sufficiently thin for observation, point defect clusters had already appeared. Experiments explained in this section were carried out by applying an additional deformation to an area which already contained point defect clusters.

In Fig. 7 (a) is before additional deformation, (b) is after deformation, and (c) illustrates the change in positions of vacancy clusters that survived the additional deformation. A narrow band between the two regions has undergone elongation deformation. Almost all the clusters in the deformed area disappeared during the deformation. This observation shows that the deformation proceeds by a mechanism that does not involve dislocations, and the atomistic reaction for carrying out deformation and for eliminating existing vacancy clusters occurs homogeneously over the deformed area. When an additional deformation is heavily localized, new clusters appeare.

Point defect clusters formed by deformation often exhibit nonuniform distribution, resembling scattered clouds in the sky. This nonuniform distribution is common to all metals examined, but is not pronounced when the fracture of the films occurred suddenly at high speed. In other words, inhomogeneity is pronounced when a film comes to fracture slowly or by intermittent deformation. A large difference between adjacent areas having no remarkable variation in deformation morphology can be understood to derive from a difference in deformation history between adjacent areas.

TEMPERATURE AND STRAIN RATE DEPENDENCE

Variation of vacancy cluster formation with temperature and deformation speed was examined for three metals, Al, Au and Cu, changing the strain rate over a wide range from 10^{-4} to 10^{8}/s.

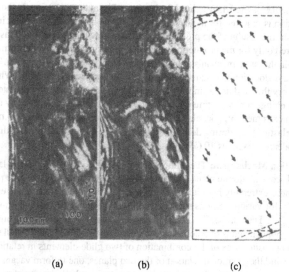

(a)   (b)   (c)

Fig. 7. Disappearance of vacancy clusters by additional deformation in gold.
(a) Vacancy clusters formed homogeneously. (b) After additional elongation.
(c) Arrows show relative movement of vacancy clusters between (a) and (b).

Details are in another paper in this proceedings [7].

The number density of SFTs saturates above a certain deformation rate, and it decreases at the slower strain rates finally reaching zero density. The strain rate at which the number density of vacancy clusters starts to decrease steeply towards the smaller strain rate is found to be proportional to vacancy mobility. This suggests that the reason for the absence of vacancy cluster formation at small strain rates is that vacancies escaped to the specimen foil surfaces during deformation. This interpretation supports the idea that vacancies are produced as vacancies, and vacancy clusters are formed from their reaction through diffusion.

High deformation speed is required for the formation of vacancy clusters, but this does not necessarily mean that deformation without dislocation and deformation to produce vacancies at high density is characteristic to high-speed deformation.

DISCUSSION

Plastic Deformation without Dislocations: Although dislocations can generate point defects, by processes such as the motion of jogs and mutual annihilation of dislocations of different signs, the generation should be heterogeneous. However, vacancy clusters produced by deformation are distributed so homogeneously that they are not affected even by the presence of specimen surfaces. SFTs might be formed directly by geometrical reaction between dislocations [5, 8], but the clusters in the present case are formed from dispersed vacancies produced by deformation. Simply, the heavily deformed area contains no dislocations, nor indicates the operation of dislocations.

The direct observation of the progress of deformation confirmed the progress of deformation by a mechanism not involving dislocations. The observation was performed under carefully adjusted electron microscopic imaging conditions under which dislocations would have been clearly observed if any existed.

Reasons for the Absence of the Dislocation Mechanism: The authors previously thought that high-speed deformation is required for the occurrence of plastic deformation without involving dislocations. However, the results of the present experiment show that deformation faster than a certain limit is required only for the production of point defect clusters. Slower deformation is thought to proceed via the same mechanism as in the case of high-speed deformation. The reason for the absence of the dislocation mechanism is thought in the specimen geometry. When a foil specimen becomes very thin and has a smooth surface, no singularity that can generate dislocations is present. When the elongation is continued, the elastic strain increases continuously and internal stress increases to an extremely high level. As a result, plastic deformation via a new mechanism starts. Actually, the elastic strain during deformation is recently detected to be more than 10 %, and the estimated internal stress exceeds 10 GPa.

Proposed Deformation Mechanism: The authors previously proposed a model of plastic deformation in which small areas of atomic planes, presumably the same planes as the glide planes of dislocations, shift under extremely high internal stress, and named these areas 'glide elements' [1]. These glide elements, for various reasons such as the inhomogeneous thermal vibration of atoms, might not extend largely. The number density of glide-elements must be extremely high for heavy deformation, and the opposite ends of a combination of two glide-elements never fail to meet each other. There are two opposite cases of the combination of two glide-elements in relation to the direction of the glide and the positional relation of the two planes, one to form vacant atomic sites and the other to produce extra-atoms. The sample is covered with a high density of glide elements joined together, producing a tremendous number of point defects and small point defect complexes.

If we simply assume geometrical symmetry, the production of pairs of glide elements, which produce vacancies and interstitials, will be equal to each other and vacancies and interstitials will be produced in equal numbers. However, production of the pair that produces interstitials is thought to face stronger resistance, and the concentration of these kinds of pairs will be less than those which produce vacancies. In such a case, the number of vacancies exceeds the number of interstitials, and this is thought to appear as vacancy clusters.

ACKNOWLEDGEMENT

This work was supported by the Ministry of Education, Culture, Sports, Science and Technology of Japan as an Academic Frontier Research Project on High-speed Plastic Deformation.

REFERENCES

1. M. Kiritani, Y. Satoh, Y. Kizuka, K. Arakawa, Y. Ogasawara, S. Arai and Y. Shimomura: Phil. Mag. Letters, 79 (1999) 797.
2. Inter. Conf. *Dislocations 2000* (NIST, Gaithersburg, 2000).
3. M. Kiritani: Rad. Eff. and Defects in Solids, 148 (1999) 233.
4. H. G. F. Wilsdorf: Mater. Sci. & Eng., 59 (1983) 1.
5. S. Kojima, Y. Satoh, H. Taoka, I. Ishida, T. Yoshiie and M. Kiritani: Phil. Mag. A59 (1989) 519.
6. M. Kiritani, M. Konno, T. Yoshiie and S. Kojima: Mater. Sci. Forum, 15-18 (1987) 181.
7. K. Yasunaga, Y. Matsukawa, M. Komatsu and M. Kiritani: in this volume.
8. M. H. Loretto, L. M. Clarebrough and R. L. Segall: Phil. Mag., 11 (1965) 459.

AUTHOR INDEX

SUBJECT INDEX

Printed in the United States
By Bookmasters